J. A. Gewert, J. Görlitzer, S. Götze, J. Looft, P. Menningen, T. Nöbel, H. Schirok, C. Wulff

Organic Synthesis Workbook

Foreword by Erick M. Carreira
Translated by William E. Russey

Further Reading from Wiley-VCH

Constable, E. C.
Metals and Ligand Reactivity
An Introduction to the Organic Chemistry of Metal Complexes
1996. 308 pp. Cloth ISBN 3-527-29278-0. Paper, ISBN 3-527-29277-2

Cornils, B./Herrmann, W. A.
Applied Homogeneous Catalysis with Organometallic Compounds
2000. 1286 pp. Paper, ISBN 3-527-29594-1

Fuhrhop, J./Penzlin, G.
Organic Synthesis
Concepts, Methods, Starting Materials
1994. 432 pp. ISBN 3-527-29074-5

Latif-Ansari, R./Qureshi, R./Latif-Qureshi, M.
Electrocyclic Reactions
Fundamentals to Research
1998. 288 pp. ISBN 3-527-29755-3

Lehn, J.-M.
Supramolecular Chemistry
Concepts and Perspectives
1998. 288 pp. Paper, ISBN 3-527-2311-6

Mulzer, J./Waldmann, H. (ed.)
Organic Synthesis Highlights III
1998. 460 pp. Paper, ISBN 3-527-29500-3

Nicolaou, K. C./Sorensen, E. J.
Classics in Total Synthesis
1994. 350 pp. Paper, ISBN 3-527-29231-4

Nogradi, M.
Stereoselective Synthesis: A Practical Approach
1995. 387 pp. Paper, ISBN 3-527-29243-8

J. A. Gewert, J. Görlitzer, S. Götze, J. Looft,
P. Menningen, T. Nöbel, H. Schirok, C. Wulff

Organic Synthesis Workbook

Foreword by Erick M. Carreira
Translated by William B. Russey

WILEY-VCH

Weinheim · New York · Chichester · Brisbane · Singapore · Toronto

J. A. Gewert
J. Görlitzer
S. Götze
J. Looft
P. Menningen
T. Nöbel
H. Schirok
C. Wulff
Institut für Organische Chemie der
Universität Göttingen
Tammannstraße 2
D-37077 Göttingen

Translated by: Dr. William E. Russey, Professor of Chemistry, Juniata College, Huntingdon, PA 16652, USA

Library of Congress Card No. applied for.

A cataloque record for this book is available from the British Library.

Die Deutsche Bibliothek – Cataloguing-in-Publication Data
A catalogue record for this publication is available from Der Deutschen Bibliothek
ISBN 3-527-30187-9

1st Edition 2000
1st Reprint 2003
2nd Reprint 2004
3rd Reprint 2006

Printed on acid-free and chlorine-free paper.

Composition: K+V Fotosatz GmbH, D-64743 Beerfelden
Printing: Strauss Offsetdruck, D-69509 Mörlenbach
Bookbinding: Wilh. Osswald & Co., D-67433 Neustadt/Weinstr.

Printed in the Federal Republic of Germany

Dedicated to our PhD supervisor
Prof. Dr. Dr. h. c. L. F. Tietze.

Foreword

"He that questioneth much shall learn much..."

F. Bacon, *Of Discourse*

At the core of synthetic chemistry is the desire to understand the relationship between molecular structure and reactivity and to harness insights into this relationship in forging reactions, reagents, or catalysts for the synthesis of novel molecules and macromolecular ensembles. The sheer breadth and scope of modern synthetic organic chemistry can seem to a newcomer a labyrinthine landscape that is difficult to navigate. For all of us life-long students of organic chemistry one of the most effective means of assimilating the dramatic advances that are continually taking place in reaction innovation is a careful study of natural products syntheses. Indeed, this endeavor can serve as a critical barometer that permits synthetic science, its advances and persistent hurdles, to be gauged. Pedagogically, the study of synthetic strategies provide a powerful means to learn reaction chemistry, and the interplay between structure and reactivity.

The last few years have seen a revolution in synthetic chemistry as intriguing discoveries at the various interfaces of the chemical sciences have engendered phenomenally new reagents and catalysts for organic synthesis, allowing innovative new tactics and strategies to be considered and implemented. Indeed, the report of innovative methodology is followed quickly by ingenious applications to molecule synthesis. The careful analysis and study of complex molecule syntheses has much to teach the curious about various aspects of chemistry that are brought into play in imaginatively crafting a workable synthetic strategy. The *Organic Synthesis Workbook* is an ideal compilation of state-of-the art modern syntheses which wonderfully showcases the latest advances in synthetic chemistry in combination with fundamentals in a question-and-answer format. The structure of the book is such that the reader can appreciate the intricacies of strategic planning, reagent tailoring, and structural analysis within the context of the individual synthetic targets. In providing highlights of synthesis from a wider range of natural products classes (alkaloids, terpenes, macrolides) the reader is given a tour through a broad range of reaction chemistry and concepts.

Moreover, because in its scope the authors have ignored international borders, the book effectively parlays the global aspect of current research in the exciting field of organic synthesis.

When I was a graduate student, I recall having two books on my shelf that were frequently consulted for study: *The Art of Problem Solving in Organic Chemistry* by Alonso and *Organic Synthesis* by Ireland. Who could forget chapter titles in the latter such as *"Where in the Carbon Skeleton Is the Thing"* or *"Stereochemistry Rears Its Ugly Head*". These books provided a source of challenging problems that kept my lab mates and I captivated and fuelled energetic discussions. Moreover, whenever intriguing problems were discussed at group meetings, many of us would jot these pearls down for future reference and as a keepsake of that exciting moment when a key concept had been made clear – the Eureka! we all yearn for. It is interesting to note how much has transpired in the science of synthesis in the intervening period by comparing and contrasting each of these with the *Organic Synthesis Workbook*. Even a cursory examination makes evident to the curious how little overlap exists – and this in the course of at most 15 years! One can think of few more powerful testimonies to the vitality of the continuously advancing field of organic synthesis. The *Organic Synthesis Workbook* promises to be to the current generation of graduate students, and even "students-for-life", what Ireland's and Alonso's books were to those of us who were graduate students in the 80's. The authors have wonderfully captured the thrill, the enjoyment, and the intellectual rigor that is so characteristic of modern synthetic organic chemistry.

Erick M. Carreira

Preface

We hope you'll find working with this problem book to be an enjoyable experience!

A wonderful tradition in the research group of Prof. L.F. Tietze at the University of Göttingen is a seminar entitled *Problems*. This seminar provides the opportunity in a relaxed, conversational setting for participants to rack their brains over natural-product syntheses presented to them in tantalizingly fragmentary form. No holds are barred when it comes to questions that might be raised!

Insofar as possible we have attempted to recreate that same atmosphere in our exercise book.

The book is directed toward advanced students of organic chemistry – graduate and undergraduate – who have a special interest in synthesis.

The subject matter is distributed over 16 mutually independent chapters in such a way as to encourage individual study, but the solitary reader is never left entirely at the mercy of his or her own ingenuity as far as solving the problems is concerned.

Each chapter begins with a brief *Introduction*, which serves mainly to acquaint the reader with the nature and origin of the current target molecule. This is followed by a double-page spread, the *Overview*, which outlines the various challenges posed in that particular chapter, setting them within the context of an overall reaction scheme. Already at this point the interested reader can begin dealing with a fresh set of puzzles drawn from an imaginative and timely total synthesis of a complex organic molecule.

Every *Overview* is followed by a detailed section entitled *Synthesis*, where each individual problem is examined, beginning with its restatement in the form of the appropriate chemical equation set against a gray background for emphasis. Each problem is characterized by an interesting gap that must be bridged: a missing starting material, product, or set of essential reagents. If reagents and reaction conditions are to be elucidated, the corresponding equation arrow will lack the customary explanatory label. Sometimes several steps are grouped together as one problem, in which case a clear indication is provided of the number of operations required.

One by one the equations are then elaborated under three headings: *Problem*, *Tips*, and *Solution*. A few cases have been supplemented with a somewhat broader *Discussion*. The beginning of each new subsection is signaled in the outer margin in such a way that the conscientious reader can easily hide with a sheet of paper the ensuing clues (i.e., the *Tips*, which become progressively more explicit), to be unveiled one at a time as the need arises. One can thus easily avoid premature exposure to too much information and defer help until such time as it becomes essential.

We wish our readers much pleasure and satisfaction as they work their way through this collection of *Problems*. Suggestions for improvement – indeed, reader responses of every sort! – will be enthusiastically welcomed.

Jan-Arne Gewert
Jochen Görlitzer
Stephen Götze
Jan Looft
Pia Menningen
Thomas Nöbel
Hartmut Schirok
and Christian Wolff

Göttingen, 2000

Contents

List of Authors

Dr. J.A. Gewert
Am Brook 25
27476 Cuxhafen
Germany

Dr. J. Görlitzer
Nikolaikirchhof 12
37073 Göttingen
Germany

Dr. S. Götze
Gut Kallenhof
Rheydter Straße 301
41464 Neuss
Germany

Dr. J. Looft
Moosbergstraße 51
64285 Darmstadt
Germany

Dr. P. Menningen
Gut Kallenhof
Rheydter Straße 301
41464 Neuss
Germany

Dr. T. Nöbel
Mainstraße 8
67117 Limburger Hof
Germany

Dr. H. Schirok
2060 Amherst Street
Palo Alto, CA 94306
USA

Dr. Ch. Wulff
Mollstraße 18
68165 Mannheim
Germany

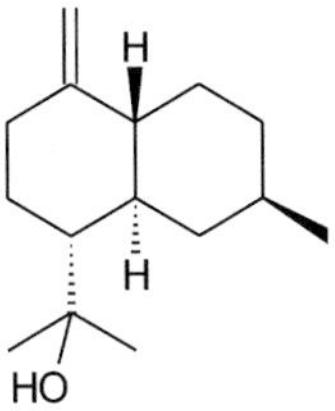

1

Veticadinol: Tietze (1998)

1.1 Introduction

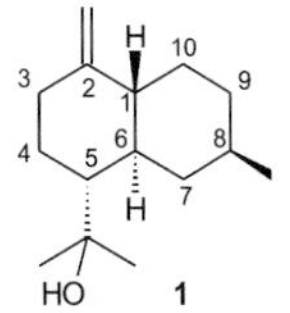

Veticadinol (**1**) is classified as a sesquiterpene. Terpenes are compounds containing two or more isoprene units, and sesquiterpenes consist of three isoprene units.

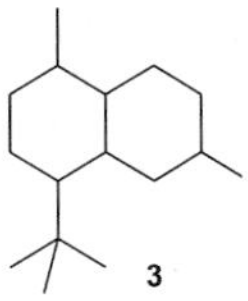

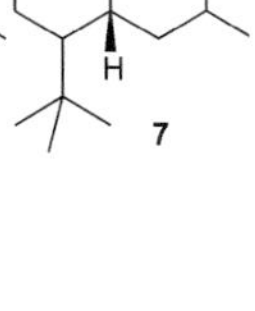

Terpenes are found primarily in higher plants, less commonly in animals. They are often isolated by steam distillation or extraction. The major commercial applications of terpenes are as fragrance and flavoring agents. Terpenes often serve in organic chemistry as versatile carriers of chiral information. For example, (–)-α-pinene (**2**) is commonly used as a ligand for chiral induction.

Sesquiterpenes that contain the 2,8-dimethyl-5-isopropyldecalin skeleton **3**, like veticadinol (**1**), are referred to as candinanes. Depending on the nature of the ring junction these are subdivided into the true candinanes **4** and bulgaranes **5**, both with a *trans* ring fusion, and the muurolanes **6** and amorphanes **7**, both containing a *cis*-fused decalin system.

Veticadinol (**1**) was isolated by *Chiurdoglu* and *Delsemme* in 1961 from Congolese vetiver oil,[1] but could not at first be obtained in pure form. Various groups attempted to synthesize a substance with the structure **1** that had been assigned to veticadinol, but each attempt led to mixtures of isomeric compounds.

The synthetic sequence presented here was the first that permitted a stereoselective preparation of veticadinol (**1**).[2]

HO 1

2

3

4 5

6 7

1.2 Overview

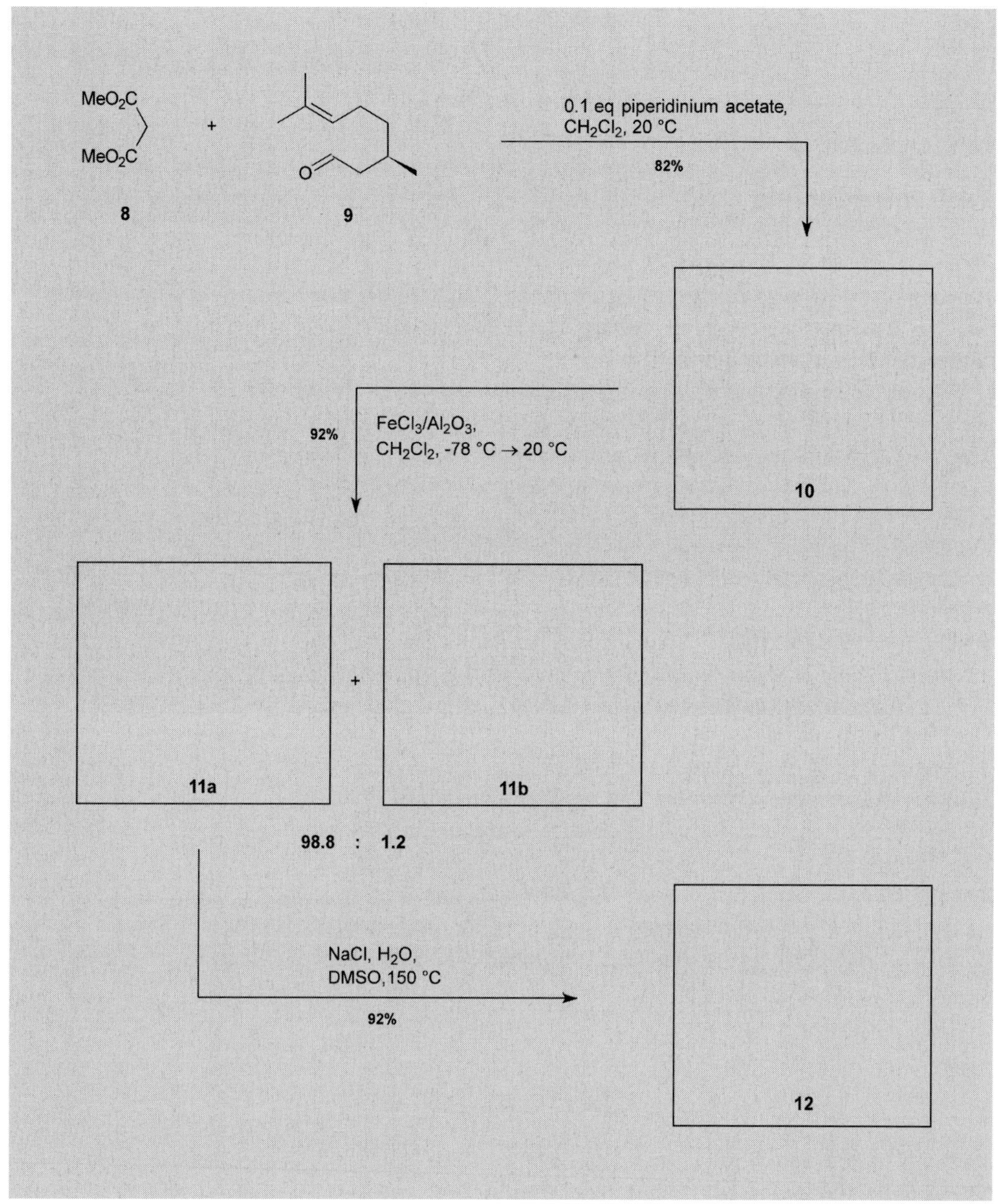
MeO2C
MeO2C
+
8
9
0.1 eq piperidinium acetate,
CH2Cl2, 20 °C
82%
10
92%
FeCl3/Al2O3,
CH2Cl2, -78 °C → 20 °C
11a
+
11b
98.8 : 1.2
NaCl, H2O,
DMSO,150 °C
92%
12

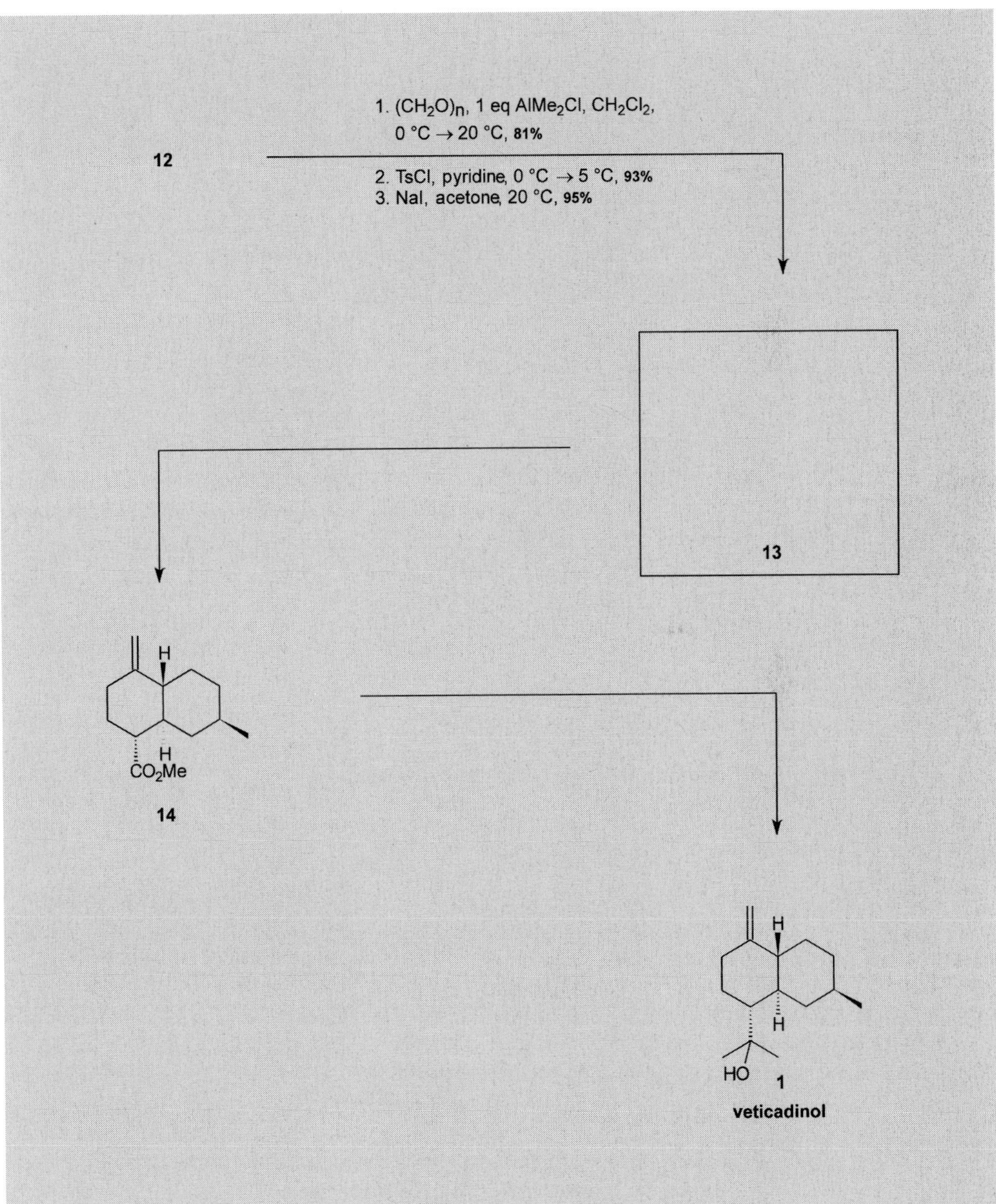

12
1. $(CH_2O)_n$, 1 eq $AlMe_2Cl$, CH_2Cl_2, 0 °C → 20 °C, 81%
2. TsCl, pyridine, 0 °C → 5 °C, 93%
3. NaI, acetone, 20 °C, 95%
13
H
H
CO_2Me
14
H
H
HO
1
veticadinol

1.3 Synthesis

Problem

MeO_2C + ... 0.1 eq piperidinium acetate, CH_2Cl_2, 20 °C, 82%

8 9 10

Tips

- Piperidinium acetate deprotonates malonic ester **8**.
- The malonic ester anion performs a nucleophilic attack on the aldehyde citronellal (**9**).
- The initially formed product eliminates water.
- The overall result is a condensation reaction.

Solution

This process represents a *Knoevenagel* condensation,[3] a reaction in which a compound with an acidic methylene group, such as dimethyl malonate (**8**), condenses with a carbonyl compound like citronellal (**9**) to give an alkene. Reaction occurs in a weakly basic or neutral medium. Catalytic amounts of piperidinium acetate suffice to deprotonate malonic ester **8**. The resulting anion **15** adds to the aldehyde citronellal (**9**) to give alkoxide **16**.

H, H, CO_2Me, CO_2Me 8

NH_2 OAc

- HOAc

H, CO_2Me, CO_2Me 15

MeO_2C MeO_2C + O 15 9

MeO_2C MeO_2C O 16

+ $H^{\oplus}$ - H_2O

MeO_2C CO_2Me 10

The latter is immediately protonated and, in the presence of an acid catalyst, it loses water to give diene **10**.

Problem

MeO$_2$C
CO$_2$Me
10

92% FeCl$_3$/Al$_2$O$_3$, CH$_2$Cl$_2$, -78 °C → 20 °C

11a + **11b**

98.8 : 1.2

Tips

- A six-membered ring is closed.
- The process occurs as a pericyclic reaction.
- A hydrogen atom migrates in a 1,5-shift.
- The location of the allylic double bond changes.
- An ene reaction takes place.

Solution

In a Lewis-acid catalyzed intramolecular ene reaction,[4] the alkene (ene) reacts in a pericyclic fashion with an enophile to form two new σ bonds with concurrent migration of the allylic double bond.

En
H
MeO$_2$C
Enophil
CO$_2$Me
10

MeO$_2$C
CO$_2$Me
11a

MeO$_2$C
CO$_2$Me
11a

MeO$_2$C
CO$_2$Me
11b

The result is two diastereomers in the ratio 98.8 to 1.2, where the desired product **11a** predominates.

The reaction occurs with *trans* selectivity, as would be expected given the chair-like nature of the transition state. Of the four possible structures, the *endo-Z-anti*-transition state **17** is especially unfa-

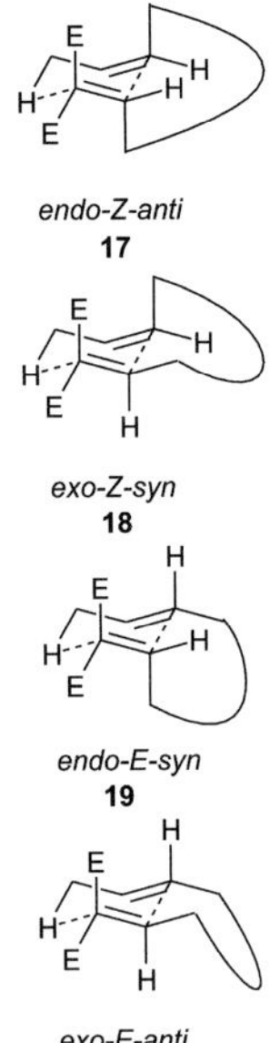

endo-Z-anti **17**

exo-Z-syn **18**

endo-E-syn **19**

exo-E-anti **20**

vorable due to pseudoaxial placement of the two ends of the chain, which also results in considerable 1,3-diaxial strain. The *exo-Z-syn* (**18**) and *endo-E-syn* (**19**) arrangements both include one chain terminus in a pseudoaxial orientation and are thus roughly equivalent in energy, but less favored than *exo-E-anti* transition state **20**, in which both chain termini occupy pseudoequatorial positions. This corresponds to the energetically most favorable reaction pathway, especially for six-membered rings.

H, CO_2Me, MeO_2C, CH_3, H

exo-E-anti pseudoequatorial **20a**

H, CH_3, MeO_2C, H, CO_2Me

exo-E-anti pseudoaxial **20b**

Two possibilities can be distinguished with compound **10** for an *exo-E-anti* transition state: the methyl group might occupy either a pseudoequatorial or a pseudoaxial position. Product **11a** arises from the energetically more favorable state **20a**, whereas **11b** is derived from transition state **20b**. It is in this way that the methyl group of citronellal (**9**) determines the ratio of **20a** to **20b**.

Discussion

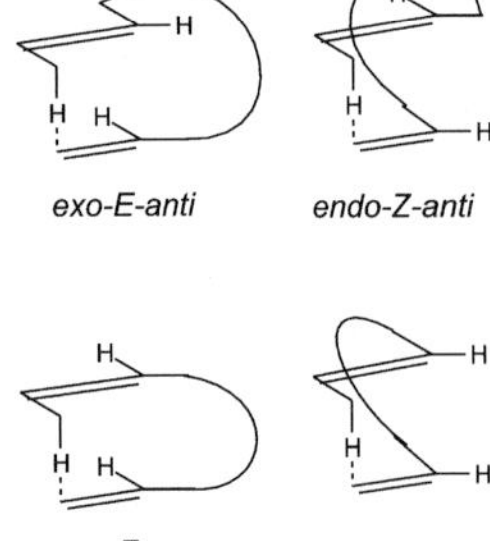

exo-E-anti *endo-Z-anti*

exo-Z-syn *endo-E-syn*

The following nomenclature rules apply to describing the transition state structures. The terms *exo/endo* refer to orientation of the chain starting from the enophile and proceeding to the ene. *Endo* means that the chain points in the direction of the ene, whereas in an *exo* form it points away from the ene. *E* and *Z* describe the geometry of the double bond of the ene. *Syn* and *anti* characterize the positions of the vinylic hydrogen atoms relative to each other.

Various reaction mechanisms are known for ene reactions. Both single-step synchronous reaction and stepwise processes involving diradicals or zwitterionic transition states have been discussed. One of the three bonds broken in the course of the reaction is a σ bond, which dictates a high activation energy relative to a *Diels-Alder* reaction (see Chapter 2). For this reason if the reaction is conducted thermally, temperatures above 100 °C are required. However, the reactivity of the enophile can be increased by addition of a Lewis acid, permitting milder reaction conditions. The Lewis acid coordinates with the electron-withdrawing group of the enophile and thus lowers the energy difference between the reacting orbitals, the LUMO of the enophile and the HOMO of the ene, thereby accelerating the reaction.[5]

Problem

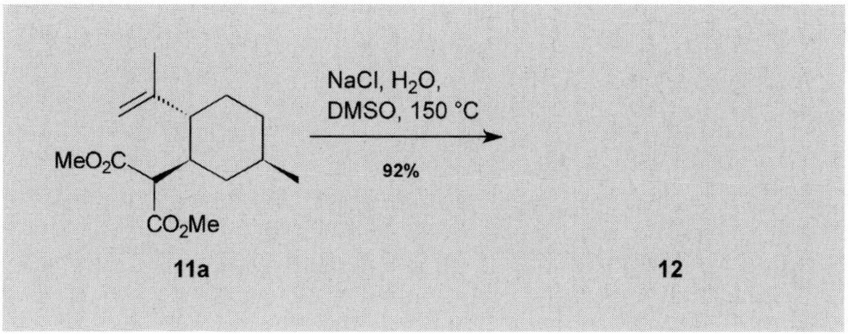

NaCl, H_2O, DMSO, 150 °C

92%

11a **12**

Tips

- A gas is evolved.
- This gas is CO_2.
- Only one equivalent of CO_2 is released.

Solution

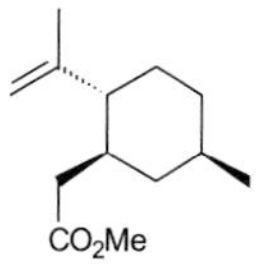

The reaction is a decarboxylation of the *Krapcho* type,[6] in the course of which one of the two ester groups of the starting malonate is removed and replaced by a proton. What results is ester **12**. The only aspect of the reaction mechanism that has been established is that it does not proceed via the free acid. *Krapcho* himself has proposed the following mechanism:

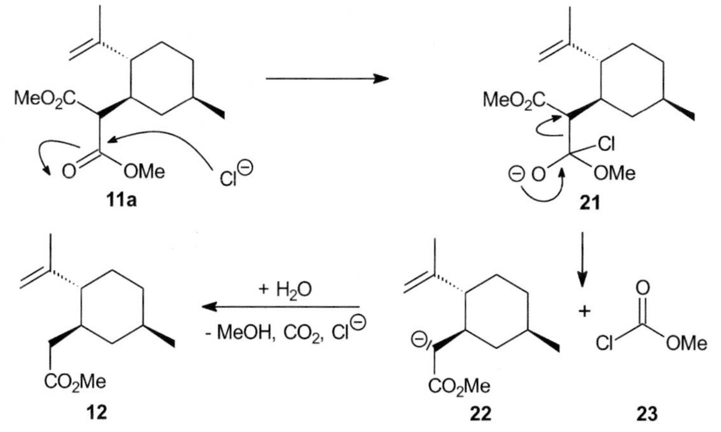

11a **21** **22** **23** **12**

+ H_2O

- MeOH, CO_2, $Cl^{\ominus}$

The chloride ion thus carries out a nucleophilic attack on one of the two carbonyl groups of malonic ester **11a**. Ester anion **22** is formed with cleavage of a carbon-carbon bond, as is chloroformic acid **23**. Ester anion **22** is subsequently protonated by water. Chloroformic acid **23** hydrolyzes immediately to chloride ion, methanol, and carbon dioxide.

Problem

CO_2Me

12

1. $(CH_2O)_n$, 1 eq $AlMe_2Cl$, CH_2Cl_2, 0 °C → 20 °C, 81%

2. TsCl, pyridine, 0 °C → 5 °C, 93%
3. N I, acetone, 20 °C, 95%

13

Tips

- The first step involves an attack on the double bond.
- The chain is extended by one hydroxymethylene group.
- This is an ene reaction with formaldehyde.
- What results is a primary alcohol, which undergoes nucleophilic substitution in the second and third steps.

Solution

HO

CO_2Me

24

I

CO_2Me

13

The first step is a carbonyl ene reaction, also known in the literature as a *Prins* reaction.[7] A Lewis acid activates formaldehyde (**25**) for attack on the double bond of **12**. This results in zwitterionic intermediate **26**, which leads to the ene product **27** in the form of a dimethylaluminum complex through 1,5-migration of a proton. This complex is unstable and spontaneously eliminates methane. Aqueous workup hydrolyzes aluminum alkoxide **28** to alcohol **24**.

The resulting homoallylic alcohol **24** is next transformed into a tosylate, which subsequently undergoes nucleophilic substitution with sodium iodide in a *Finkelstein* reaction to give compound **13**.

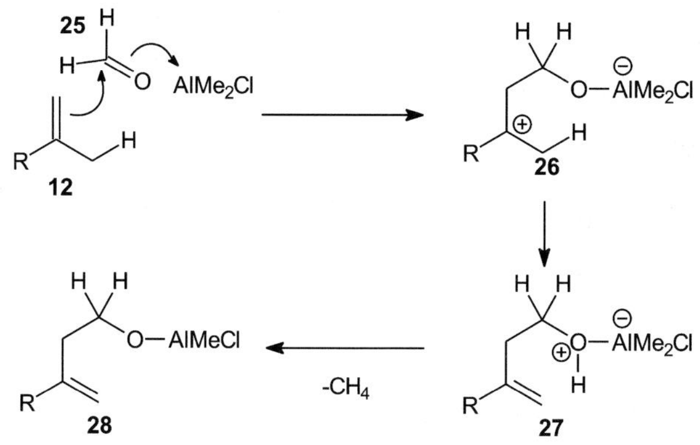

Problem

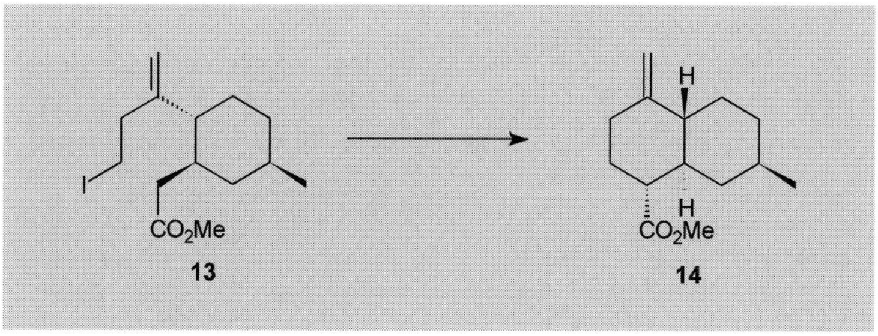

- A strong base is introduced.
- Compound **13** has acidic hydrogen atoms on a carbon α to an ester function.

Tips

Solution

This is an intramolecular alkylation, through which the second six-membered ring is closed, producing the candinane skeleton. The reagent employed is the strong base lithium diisopropylamide (LDA). This generates ester enolate **13a**, which carries out an intramolecular nucleophilic attack on the carbon bearing iodine, leading to product **14**.

If one assumes a chair-like transition state, ring closure of **13a** must generate product **14**, whereas **13b** leads to the diastereomer **29**. Due to a pseudoequatorial placement of the ester function, **13a** is energetically preferred over the pseudoaxial form **13b**, so that ring closure produces exclusively product **14**.
LDA, THF, –78 °C → 20 °C, 92%.

MeO H O⊖ 13a → MeO₂C H 14

MeO O⊖ H 13b → CO₂Me H H 29

Problem

14 → 1 veticadinol

Tips

- A reagent is introduced in excess.
- Two equivalents of an organometallic compound are added.
- A magnesium compound is employed.

Solution

The process here is a double *Grignard* reaction of methylmagnesium iodide with the ester function of **14**.[8] Addition of the first equivalent of *Grignard* reagent leads to the metallated hemiacetal **31**, which decomposes to ketone **32**. The ketone is then attacked nucleophilically once again by the magnesium species and transformed into the magnesium alkoxide of a tertiary alcohol. Aqueous workup leads to the target molecule veticadinol (**1**).

It is assumed that addition of *Grignard* reagents to carbonyl compounds results in cyclic transitions states like **30**, requiring two equivalents of organomagnesium compound. *Grignard* reagents are characterized by participation in the *Schlenk* equilibrium,[8] which means that in addition to methylmagnesium iodide, contributions to the cyclic transition states may be made by dimethylmagnesium and magnesium iodide.

2 MeMgI ⇌ Me—Mg(I)₂Mg—Me ⇌ $Me_2Mg + MgI_2$ ⇌ Me—Mg(Me)(I)MgI

Schlenk-equilibrium

3 eq. MeMgI, Et_2O, RT, 77%.

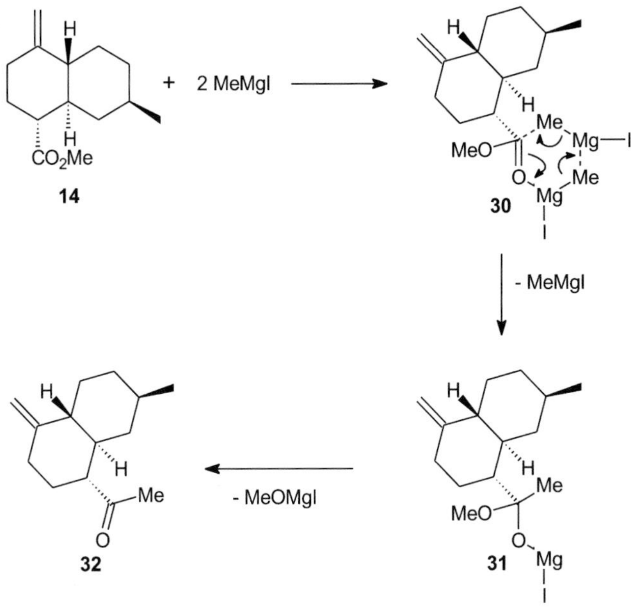

1.4 Summary

Veticadinol (**1**) was prepared enantiomerically pure through a linear eight-step synthesis with an overall yield of 36%. Starting with dimethyl malonate (**8**) and citronellal (**9**), a *Knoevenagel* reaction is used to prepare the starting material for a subsequent ene reaction. The methyl group in the terpene citronellal (**9**) directs the *trans*-selective ene reaction in such a way that the desired diastereomer **11a** is obtained with a selectivity of 98.8 to 1.2. Decarboxylation, a *Prins* reaction, and conversion into an iodide provides compound **13**, the starting material for the final key step. Alkylative cyclization of **13** selectively generates the candinane skeleton. This total synthesis is completed by a double *Grignard* reaction that leads to veticadinol (**1**).

HO 1

13

CO_2Me

MeO_2C

MeO_2C

+

O

8

9

MeO_2C

CO_2Me

11a

1.5 References

1 G. Chiurdoglu, A. Delsemme, *Bul. Soc. Chim. Belg.* **1961**, *70*, 5.

2 L.F. Tietze, U. Beifuß, J. Antel, G.M. Sheldrick, *Angew. Chem. Int. Ed. Engl.* **1988**, *100*, 703.

3 L.F. Tietze, U. Beifuß in *Comprehensive Organic Synthesis, Vol. 2*, (Ed.: B. M. Trost), Pergamon, Oxford 1991, p. 341.

4 L.F. Tietze, U. Beifuß, *Angew. Chem. Int. Ed. Engl.* **1986**, *25*, 1042; L.F. Tietze, U. Beifuß, *Tetrahedron Lett.* **1986**, *27*, 1767.

5 I. Fleming, *Frontier Orbitals and Organic Chemical Reactions*, VCH-Verlagsgesellschaft, Weinheim 1990, p. 188.

6 A.P. Krapcho, *Synthesis* **1982**, 805.

7 B.B. Snider, D. J. Rodini, T. C. Kirk, R. Cordova, *J. Am. Chem. Soc.* **1982**, *104*, 555.

8 K. Nützel in *Methoden der organischen Chemie (Houben-Weyl), Vol. XIII/2a*, Georg Thieme Verlag, (Ed.: E. Müller), Stuttgart 1973, p. 261.

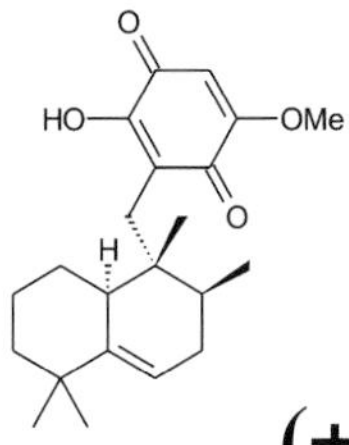

2

(±)-Mamanuthaquinone: Danishefsky (1994)

2.1 Introduction

(–)-Mamanuthaquinone (**1**) was identified in 1991 as a secondary metabolite of a marine sponge.[1] Its name is derived from the site of the organism's discovery, the island Mamanutha near the Fiji Islands. The purple moss-like sponge is classified as *Fasciospongia* sp.

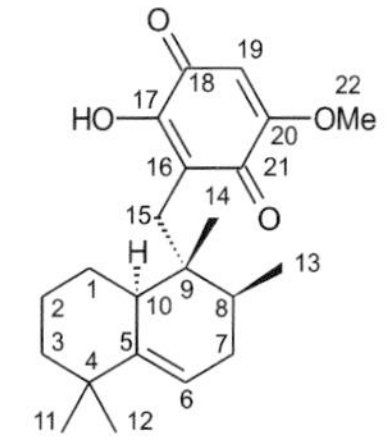

(–)-Mamanuthaquinone (**1**) is a combination of a sesquiterpene and a quinone. Its skeleton, with both terpene and quinone or hydroquinone portions, is not unusual for a natural product derived from a sponge.[2] An example is avarol (**2**), containing a hydroquinone ring and the terpene skeleton of drimane.[3]

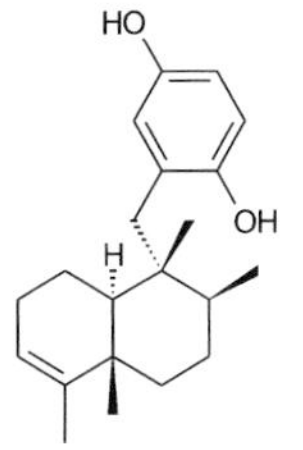

The absolute stereochemistry of (–)-mamanuthaquinone (**1**) was established through degradation reactions and subsequent comparison with known substances.[1]

The sesquiterpene quinones and hydroquinones display a wide range of biological activities.[4] Some of the compounds are cytotoxic, whereas others show antimicrobial characteristics. Avarol (**2**) has been the subject of investigation as a consequence of its ability to inhibit reverse transcriptase, but no clinical value (as an anti-AIDS agent, for example) has yet been established.[3]

Antitumor activity with respect to colon cancer has been demonstrated for (-)-mamanuthaquinone (**1**).[1] This in itself makes a synthesis of the substance of considerable interest. The synthesis devised by *Danishefsky* provides racemic mamanuthaquinone **1** in 14 steps.[5]

1

2

2.2 Overview

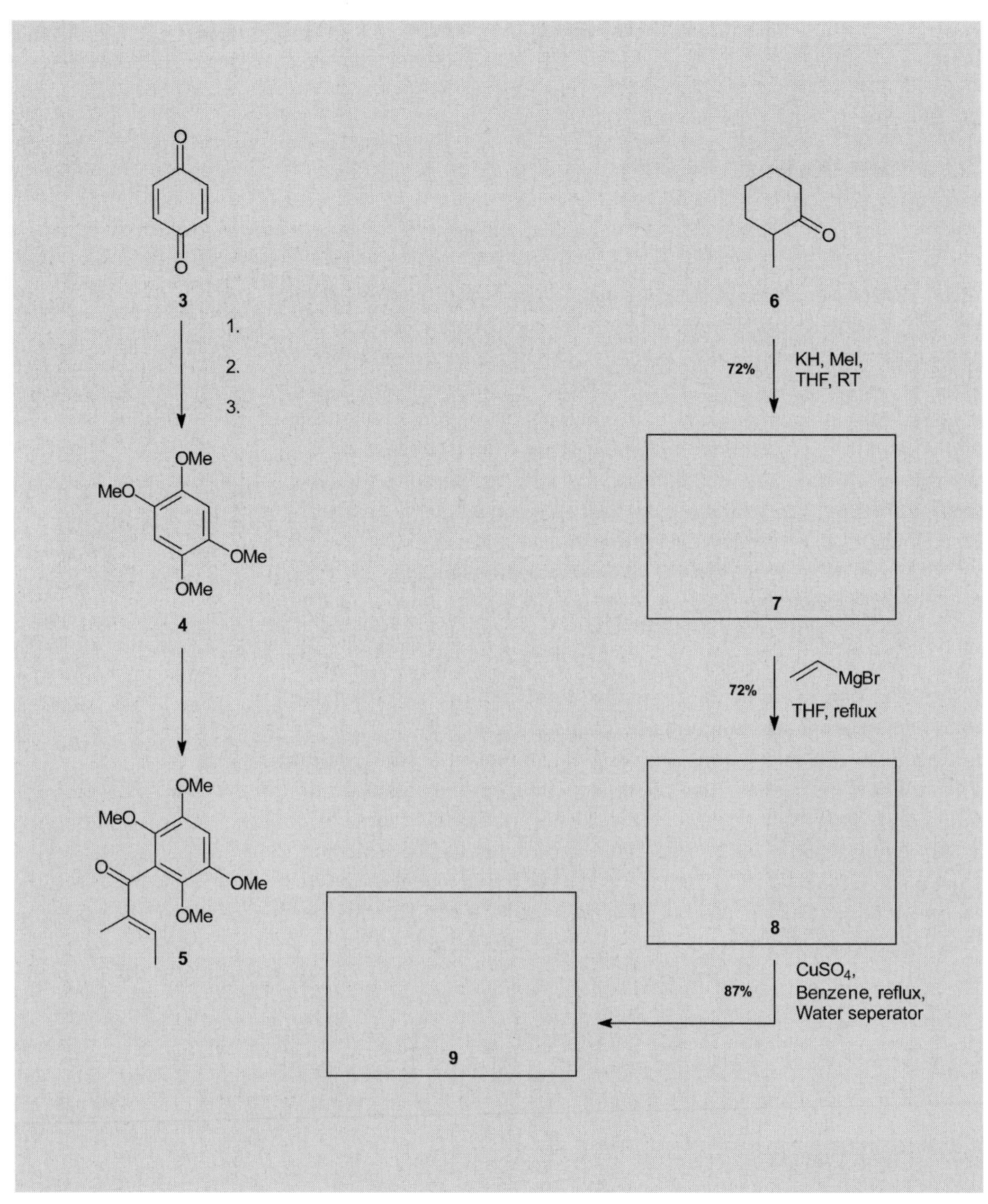
O
O
3
1.
2.
3.
OMe
MeO
OMe
OMe
4
OMe
MeO
O
OMe
OMe
5
O
6
72%
KH, MeI,
THF, RT
7
72%
MgBr
THF, reflux
8
87%
CuSO4,
Benzene, reflux,
Water seperator
9

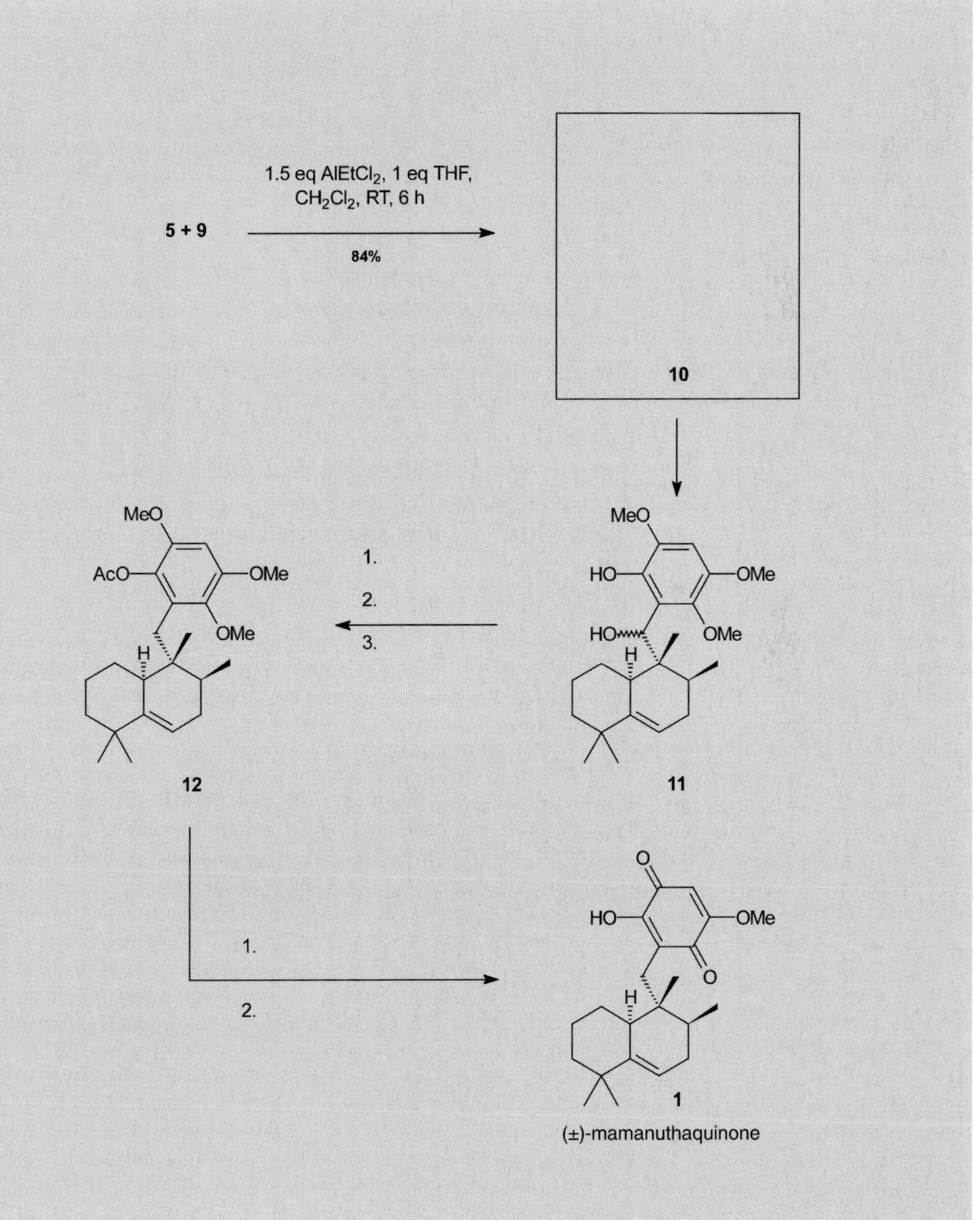
1.5 eq AlEtCl2, 1 eq THF,
CH2Cl2, RT, 6 h
5 + 9
84%
10
MeO
AcO
OMe
OMe
H
12
1.
2.
3.
MeO
HO
OMe
HO
OMe
H
11
1.
2.
O
HO
OMe
O
H
1
(±)-mamanuthaquinone

2.3 Synthesis

Problem

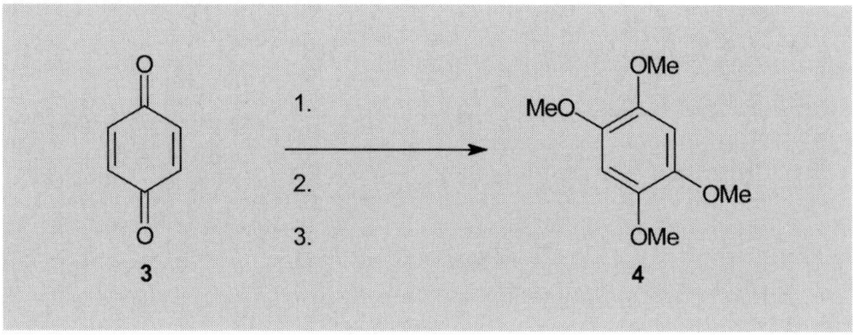

Tips

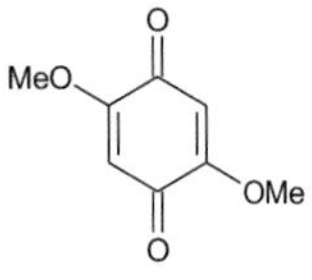

- Aromatic species appear in the reaction sequence only after all the heteroatoms have been introduced.
- The first step is a double 1,4-addition to *p*-benzoquinone (**3**).
- Two equivalents of methanol are added in the first step.
- The second reaction transforms quinone **13**, prepared in the first step, into a hydroquinone. Is this an oxidation or a reduction?
- The reducing agent utilized is converted in a redox process into hydrogensulfate.
- The third step is a double methylation.

Solution

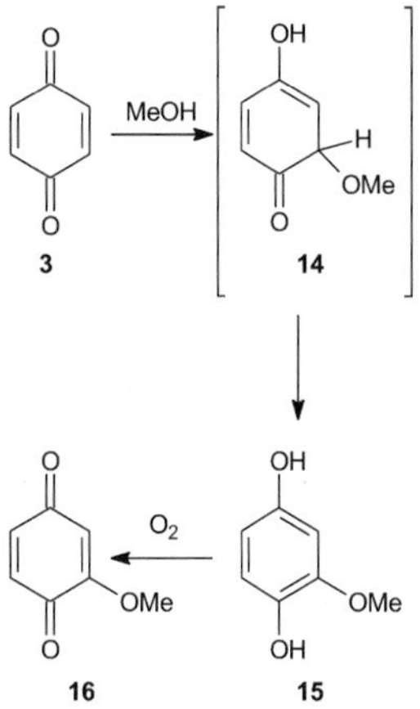

In the first step, methanol is subject to a 1,4-addition to *p*-benzoquinone (**3**).The resulting 1,4-adduct (**14**) rearranges spontaneously to hydroquinone **15**, which is oxidized by atmospheric oxygen and residual benzoquinone into a new benzoquinone **16**. A second 1,4-addition leads to benzoquinone **13**.

The latter is reduced through a second reaction with sodium hydrogensulfite as reducing agent, and in a third reaction the resulting hydroquinone is methylated at both phenol groups with dimethyl sulfate. Extra sodium hydrogensulfite is introduced during this methylation in order to avoid oxidation of the sensitive hydroquinone.[6]

1. $ZnCl_2$, MeOH, reflux, 64%.
2. 2.5 eq. $NaHSO_3$, H_2O, reflux, 78%.
3. 6.3 eq. Me_2SO_4, 0.2 eq. $NaHSO_3$, 6.2 eq. NaOH, EtOH/H_2O, 80 °C, 86%.

Problem

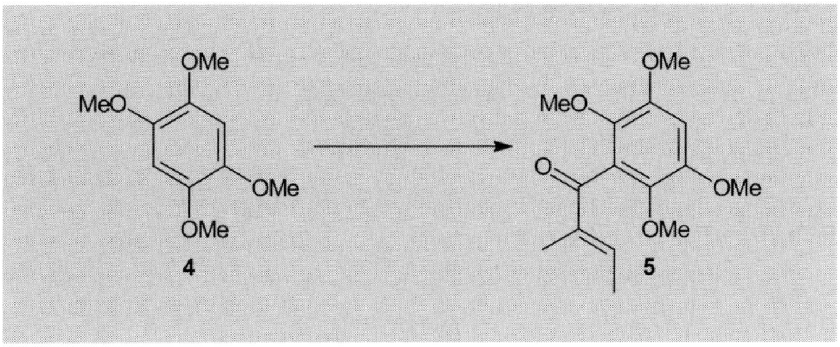

Tips

- A *Friedel-Crafts* acylation is not utilized.
- Aromatic species **4** is first lithiated.
- A special amide serves as the acylating reagent.
- Amides of this type are also used to transform carboxylic acids into aldehydes.
- The *Weinreb* amide **17** of tiglic acid is employed.

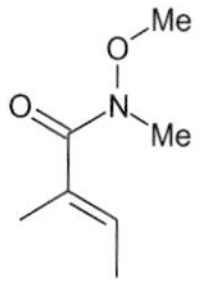

Solution

Methoxy-substituted aromatic compound **4** is lithiated metalation with Buli in THF, a step in which it proves useful to include lithium chloride. Because of the greater basicity of *n*-butyllithium relative to **4**, direct metallation is in fact possible thermodynamically, but *n*-butyllithium is generally present in solution as a tetramer, and this reduces its reactivity.[7] Addition of lithium chloride destroys these aggregates, and that eliminates the kinetic inhibition. Lithiated aromatic species **18** is further stabilized through chelate formation between lithium and the *ortho*-methoxy groups (*ortho* effect).[8]

The *N*-methoxy-*N*-methylamide of tiglic acid (**17**) is used as an acylating agent in a procedure developed by *Weinreb*.[9] Lithiated aromatic species **18** attacks *Weinreb* amide **17** with formation of the chelate **19**, which is hydrolyzed to ketone **5**. Use of *Weinreb* amide **17** circumvents the primary threat here: multiple addition and formation of a tertiary alcohol. Since complex **19** decomposes only in the course of workup, the ketone **5** itself is protected against further nucleophilic attack.[10]

^{n}BuLi, LiCl, THF, 0 °C → RT; **17**, 77%.

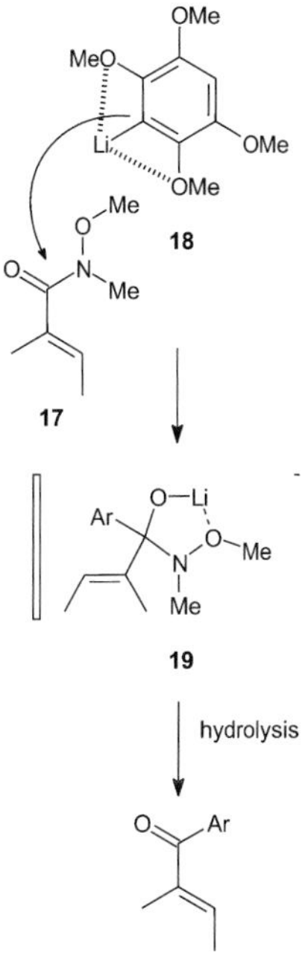

Problem

KH, MeI,
THF, RT
72%
6 → 7

Tips

- Potassium hydride produces an enolate by deprotonation.
- Which regioisomer of the enolate will form, the thermodynamic or the inetic one?
- The enolate is alkylated at carbon with methyl iodide.

Solution

Deprotonation at room temperature leads to the thermodynamically more stable enolate **20**. Treatment with methyl iodide produces cyclohexanone **7** doubly methylated in the 2-position (see Chapter 12).

20

Problem

MgBr
THF, reflux
72%
7 → 8

Tips

- The carbonyl group of starting material **7** is subjected to nucleophilic attack by a *Grignard* reagent.
- The result is a tertiary alcohol.

Solution

Nucleophilic attack by the vinyl *Grignard* reagent leads to tertiary alcohol **8**.[11] *Grignard* reagents react with formaldehyde to give primary alcohols and with other aldehydes to give secondary alcohols, whereas ketones are transformed into tertiary alcohols.

Problem

CuSO$_4$

Benzene, reflux, Water separator

87%

8 9

Tips

- Copper sulfate acts as an acidic heterogeneous catalyst.
- The driving force in the reaction is creation of a conjugated system.

Solution

This reaction is an dehydration acid-catalyzed.[12] The hexaaquocopper cation behaves as a weak cationic acid in copper-salt solution.[13] Protonation of the hydroxy group produces an oxonium ion that decomposes unimolecularly into carbocation **21** and water. Water is removed from the reaction equilibrium by means of a water-separating device. Carbocation **21** eliminates an -proton with formation of the energetically favorable conjugated diene **9**.

The great stability of tertiary carbocations is the reason why alcohols like **8** undergo unimolecular elimination especially readily.

21

Problem

OMe, MeO, O, OMe, OMe

9 + 5

1.5 eq AlEtCl$_2$, 1 eq THF,

CH$_2$Cl$_2$, RT

84%

10

Tips

- Two new C-C bonds are generated.
- The desired reaction is a [4+2]-cycloaddition.
- The system EtAlCl$_2$/THF serves as a Lewis acid and coordinates with the carbonyl oxygen atom.

- This is a *Diels-Alder* reaction with the usual demand for electrons.
- Compound **5** is the dienophile, **9** is the diene.

Solution

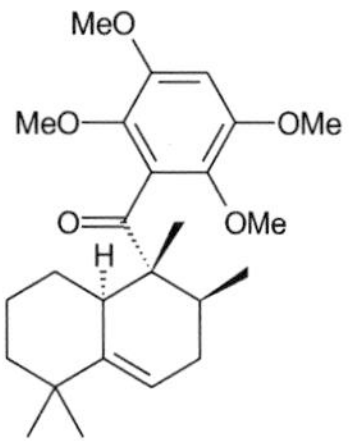

MeO MeO OMe OMe O H

10

The *Diels-Alder* reaction leads exclusively to *exo* product **10**. Steric effects are responsible for this *exo* selectivity. In a hypothetical *endo* transition state **22**, a methyl group on the quaternary carbon atom of the diene interferes with the aromatic residue of the dienophile. For this reason the reaction does not proceed via this transition state to the *endo* product **24**. Similar steric inhibition is absent in the *exo* transition state **23**.

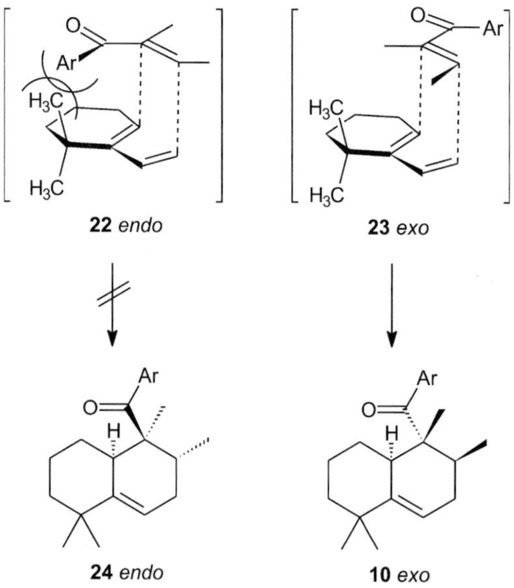

22 *endo* **23** *exo*

24 *endo* **10** *exo*

$EtAlCl_2$/THF added to the system serves as a Lewis acid that catalyzes the *Diels-Alder* reaction.[14] Introduction of THF is useful as a way of diminishing the reactivity of ethylaluminum chloride, thus making it suitable for the reaction.

Discussion

The reactivities of the diene and dienophile in a *Diels-Alder* reaction are highly dependent on their electronic structures.[15] In the case of a *Diels-Alder* reaction with normal electron demand, the dienophile is substituted with an electron acceptor *Z*, whereas the dienophile carries an electron donor *X*. The reaction in question follows this pattern. Increased reactivity in such a case can be rationalized with the frontier orbital theory of *Fukui* and *Houk*, according to which the energy difference between the (HOMO) of the diene and the lowest unoccupied molecular orbital (LUMO) of the dienophile is reduced in a favorable way by the substituents.

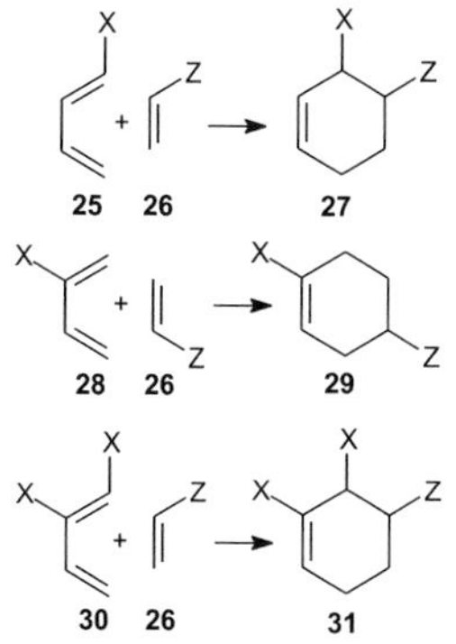

X Z 25 26 27 X Z 28 26 29 X Z 30 26 31

The electronic structures of the diene and dienophile also influence the regioselectivity of the reaction. The regiochemistry of *Diels-Alder* products **27**, **29**, and **31**, derived from the variously substituted dienes **25**, **28**, and **30**, is illustrated by the scheme in the margin.

The catalytic activity of Lewis acids in this reaction can also be interpreted on the basis of frontier orbital theory. A Lewis acid coordinates with the electron-withdrawing group on the dienophile (in this case $EtAlCl_2$ with the carbonyl group), thereby further lowering the energy of the LUMO. This in turn leads to an energy advantage and thus acceleration of the reaction.

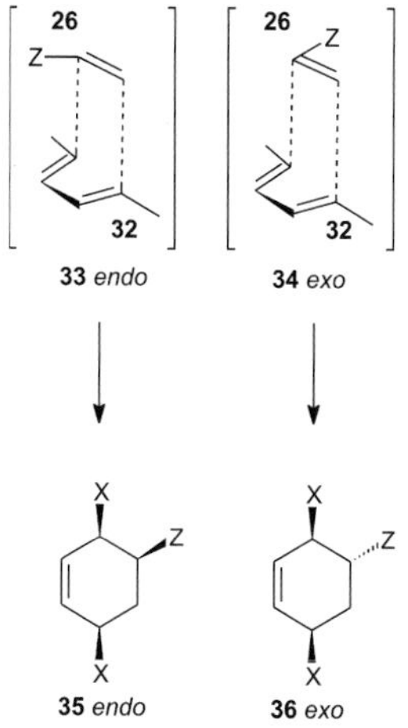

For a given regiochemistry of the *Diels-Alder* reaction there are still two orientations that might be envisioned in the case of an unsymmetrically substituted dienophile **26**. Substituent *Z* (with the highest priority) in dienophile **26** might be arranged in transition state **33** so that it is pointed toward diene **32**, which leads to the *endo* product. Alternatively, substituent *Z* might be directed away from diene **32** as in transition state **34**, leading to *exo* product **36**.[16] The *endo/exo* ratio thus reflects the simple diastereoselectivity of a *Diels-Alder* reaction.

33 *endo* **34** *exo*

35 *endo* **36** *exo*

Often one finds that *Diels-Alder* reactions lead primarily to the formation of *endo* products. This is usually explained on the basis of a secondary-overlap effect involving stabilizing interactions between orbitals not directly engaged in bond formation.[17] Nevertheless, calculations show that the energy advantage gained in energy through secondary overlap effects is minimal. The resulting preference for an *endo* transition state can easily be overcome by steric effects.[18]

Problem

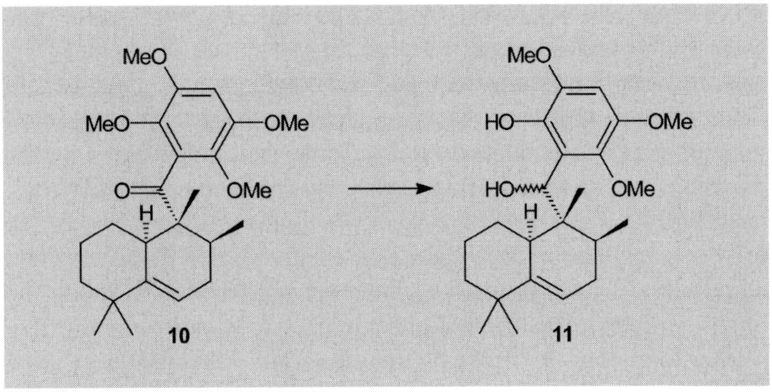

Tips

- A single step accomplishes both cleavage of an ether and reduction of the carbonyl group.
- The methyl ether is cleaved nucleophilically by hydride anion.

Solution

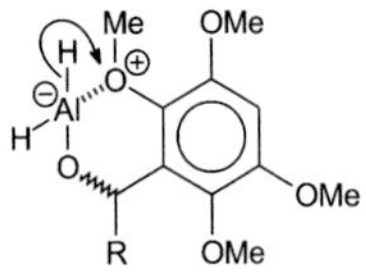

The keto function in compound **10** is reduced with lithium aluminum hydride in THF to a secondary alcohol. In the course of this reaction one of the methoxy groups in the *ortho*-position is also cleaved. It appears reasonable to explain this by an *ortho* effect: the alcohol group forms an intermediate alkoxyaluminum hydride complex **37** that coordinates with one of the methoxy groups, which is thereby activated toward nucleophilic attack by hydride. A chelate complex protects the product from cleavage of the second *ortho*-methoxy group.

$LiAlH_4$, THF, reflux.

Problem

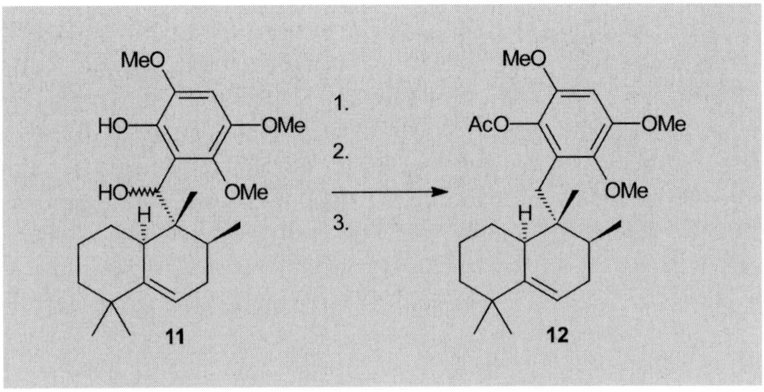

Tips

- The first step is a protecting-group operation at the phenolic and benzylic hydroxy groups.

- The protected benzylic hydroxy group is removed from the molecule in a second step.
- A *Birch* reduction of the benzylic acetoxy group is carried out in the second step.
- The phenolic hydroxy group is deprotected in the course of the *Birch* reduction, so it must again be protected.

Solution

The two free hydroxy groups are first protected with acetic anhydride. In a second step the acetyl group is reductively cleaved by a *Birch* reduction with lithium in liquid ammonia.[19] Lithium dissolves in the ammonia with the formation of solvated electrons. Stepwise electron transfer to the aromatic species (a SET process) leads first to a radical anion, which stabilizes itself as benzylic radical **38** with loss of the oxygen substituent. A second SET process generates a benzylic anion, which is neutralized with ammonium chloride acting as a proton source (see Chapter 12).

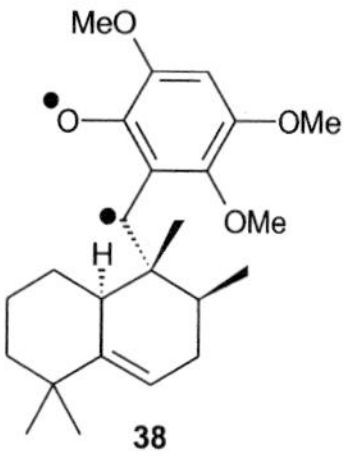

MeO
•O
OMe
OMe
H
38

Since the phenolic hydroxy group is deacetylated during the *Birch* reduction it must once again be acetylated in a third step.

1. Ac_2O, NEt_3, DMAP (cat.), CH_2Cl_2, RT.
2. 10 eq. Li, NH_3(l), –78 °C; NaOBz, NH_4Cl.
3. Ac_2O, NEt_3, DMAP (cat.), CH_2Cl_2, RT, 55% over four steps.

Problem

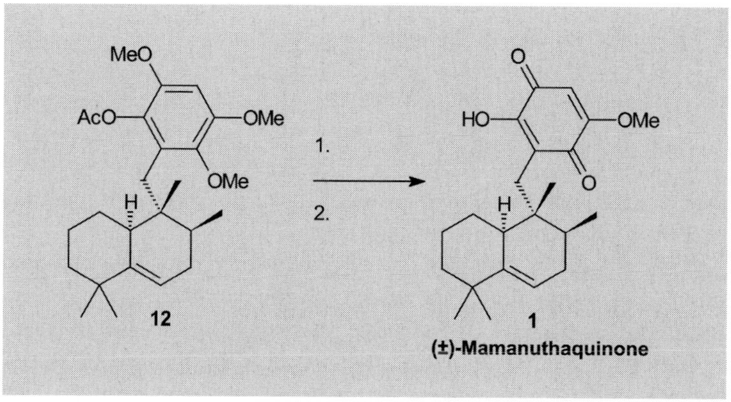

MeO
AcO
OMe
OMe
H
12
1.
2.
O
HO
OMe
O
H
1
(±)-Mamanuthaquinone

Tips

- In what sequence should transformation into a *p*-quinone and hydrolysis of the acetyl group be carried out?
- The first reaction converts protected hydroquinone system **12** into a *p*-quinone system. Should reducing or oxidizing conditions be selected?
- In the course of oxidation to a *p*-quinone the *p*-methoxy groups are demethylated.

Solution

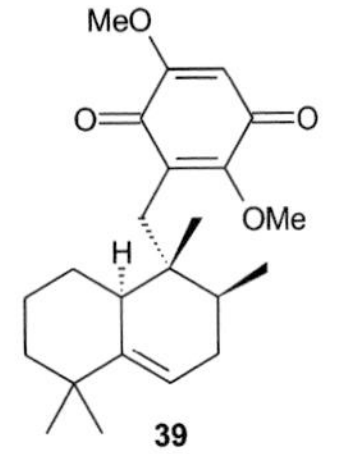

MeO O O OMe H 39

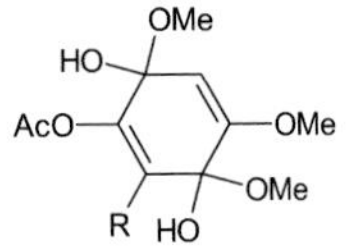

OMe HO AcO OMe OMe R HO 40

Methylated hydroquinone **12** is oxidized with the strong oxidizing agent cerium(IV) ammonium nitrate (CAN) to a quinone.[20] Protection of the phenol group by acetylation is useful here as a technique for controlling the regiochemistry of oxidation. In the absence of the acetyl group there might arise the doubly methoxy substituted *p*-quinone **39**. Mechanistically, oxidative demethylation with cerium(VI) ammonium nitrate probably proceeds via hemiacetal **40**.[21] Following oxidation, the acetyl group is hydrolyzed under basic conditions with potassium carbonate.

1. 2.5 eq. $(NH_4)_2Ce(NO_3)_6$, CH_3CN/H_2O, RT.
2. K_2CO_3, MeOH, RT, 85% over two steps.

2.4 Summary

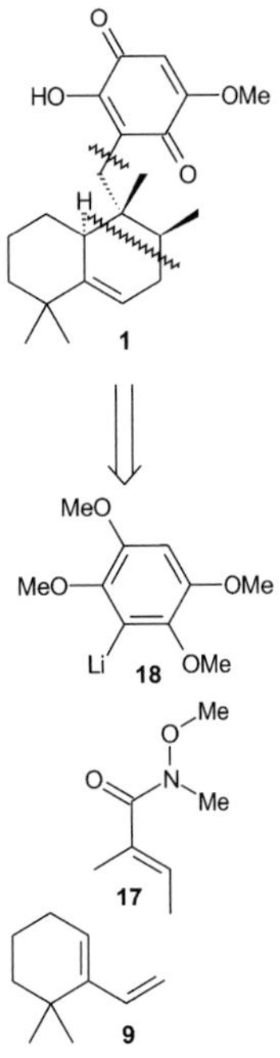

O HO OMe O H 1

MeO MeO OMe Li 18 OMe

Me O O N Me 17

9

This synthesis by *Danishefsky* provides racemic mamanuthaquinone (**1**) in 14 steps, with an overall yield of 13% based on *p*-benzoquinone.

Retrosynthetic analysis shows that mamanuthaquinone is composed of a sesquiterpene and a quinone. Quinone building block **18** is constructed using a double 1,4-addition of methanol to *p*-benzoquinone. This is followed by acylation with the C_5 building block **17**, taking advantage of a *Weinreb* amide.

The key step in the convergent synthesis is a *Diels-Alder* reaction. Diene **9** is accessible in three steps from 2-methylcyclohexanone **6**. The dienophile is an acylated *p*-benzoquinone. Surprisingly, this *Diels-Alder* reaction is completely *exo* selective and introduces the sesquiterpene skeleton of the target compound with the correct stereochemistry in a single step.

The subsequent six steps are necessary for removing a carbonyl group and transforming the hydroquinone component regioselectively into a quinone. *Danishefsky* succeeded in coupling the two operations in an elegant way. The sequence of lithium aluminum hydride and *Birch* reductions for removing the carbonyl group also accomplishes selective demethylation of one methoxy group. This selective removal of a protecting group thus ensures the correct regiochemistry in oxidative methylation to a quinone.

2.5 References

1 J.C. Swersey, L.R. Barrows, C.M. Ireland, *Tetrahedron Lett.* **1991**, *32*, 6687.

2 L. Minale in *Marine Natural Products: Chemical and Biological Perspectives Vol. 1*, (Ed.: P. J. Scheuer), Academic Press, New York 1978, Chap. 4.

3 J. Rodríguez, E. Quiñoá, R. Riguera, B.M. Peters, L.M. Abrell, P. Crews, *Tetrahedron* **1992**, *48*, 6667.

4 M.-L. Kondracki, M. Guyot, *Tetrahedron* **1989**, *45*, 1995; H.S. Radeke, C.A. Digits, S.D. Bruner, M.L. Snapper, *J. Org. Chem.* **1997**, *62*, 2823.

5 T. Yoon, S.J. Danishefsky, S. de Gala, *Angew. Chem. Int. Ed. Engl.* **1994**, *33*, 853.

6 F. Benington, R.D. Morin, L.C. Clark, Jr, *J. Org. Chem.* **1955**, *20*, 102.

7 C. Elschenbroich, A. Salzer, *Organometallics: A concise introduction*, Teubner, Stuttgart 1993.

8 P.R. Jenkins, *Organometallic reagents in chemistry*, Wiley-VCH, Weinheim 1995.

9 S. Nahm, S. M. Weinreb, *Tetrahedron Lett.* **1981**, *22*, 3815.

10 In a similar way aldehyde can be selectively protected in the presence of ketones: M.T. Reetz, B. Wenderoth, R. Peter, *J. Chem. Soc., Chem. Commun.* **1983**, 406.

11 S.P. Tanis, Y.M. Abdallah, *Synth. Commun.* **1986**, 251; *Methoden der Organischen Chemie (Houben-Weyl), Vol. XIII/2a* (Ed.: E. Müller), Georg Thieme Verlag, Stuttgart 1973, p. 86.

12 D.M. Hollinshead, S.C. Howell, S.V. Ley, M. Mahon, N.M. Ratcliffe, *J. Chem. Soc. Perkin Trans. I* **1983**, 1579.

13 A.F. Holleman, E. Wiberg, *Lehrbuch der Anorganischen Chemie*, Walter de Gruyter, Berlin 1985, p. 878.

14 L.F. Tietze, C. Schneider, *Synlett* **1992**, 755.

15 I. Fleming, *Frontier Orbitals and Organic Chemical Reactions*, VCH Verlagsgesellschaft, Weinheim 1990.

16 Die Nomenklatur folgt den *Cahn-Ingold-Prelog*-Regeln. See.: L.F. Tietze, G. Kettschau in: *Topics in Current Chemistry, Vol. 189*, Springer-Verlag, Berlin 1997, p. 1.

17 R.B. Woodward, T.J. Katz, *Tetrahedron* **1959**, *5*, 70.

18 J.G. Martin, R.K. Hill, *Chem. Rev.* **1961**, *61*, 537; Y. Kobuke, T. Fueno, J. Furukawa, *J. Am. Chem. Soc.* **1970**, *92*, 6548.

19 S. Sinclair, W.L. Jorgensen, *J. Org. Chem.* **1994**, *59*, 762.

20 A. Fischer, G. Henderson, *Synthesis* **1985**, 641.

21 P. Jacob III., P.S. Callery, A.T. Shulgin, N. Castagnoli Jr., *J. Org. Chem.* **1976**, *41*, 3627.

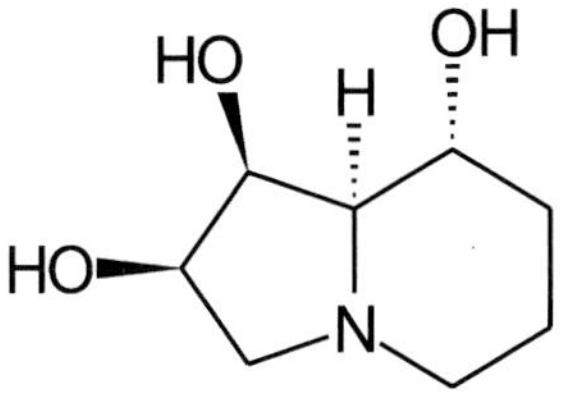

3

(–)-Swainsonine: Pearson (1996)

3.1 Introduction

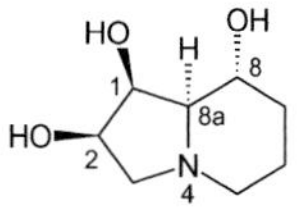

"Locoism" is the name applied to a poisoning condition that arises with grazing animals and leads to weight loss and problems with locomotion. Its symptoms resemble those of the inherited disease mannosidosis. The problem appears subsequent to ingestion of legumes of the genus *Astragalus*,[1] *Oxytropis*,[2] and *Swainsona*. Like the phytopathogenic fungi of the genus *Rhizoctonia* and *Metarhizium*,[3] these plants contain as an active ingredient the indolizidine alkaloid swainsonine (**1**).[4] The compound was first isolated in 1973 from *Rhizoctonia leguminicola* by *Broquist*[5] and later (1979) by *Colegate* from legumes of the genus *Swainsona*.[6] Structural analysis showed it to be a 1,2,8-trihydroxyoctahydroindolizine[6] with the absolute configuration 1*S*,2*R*,8*R*,8a*R*.[7]

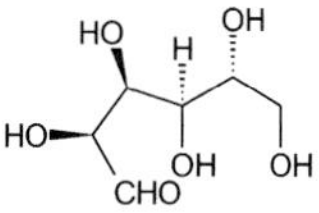

1

Since mannosidose results from a genetically derived absence of lysosomal α-mannosidase that leads to an accumulation of mannose-rich oligosaccharides, the effects of swainsonine (**1**) on this enzyme were investigated. The alkaloid, which bears a structural resemblance to the open-chain form of D-mannose (**2**), was in fact shown to be a reversible inhibitor of lysosomal α-mannosidase.[8]

2

Because of its significant biological activity as an azasugar analog of mannose (**2**), swainsonine (**1**) is of considerable synthetic interest. It has been observed that swainsonine (**1**) displays inhibitory characteristics with respect to tumor growth and metastasis, and it was the first glycoprotein metabolism inhibitor selected for clinical trials as an anticancer drug.[9] Its high cost initially ruled out commercial use of the substance. The first four total syntheses were reported in 1984,[10] and since then the number of such syntheses has multiplied rapidly.[11] The shortest synthesis, that published in 1990 by *Pearson*,[12] is not suited to large-scale implementation. By contrast, the approach described below opens the way to multi-gram quantities of swainsonine (**1**).

3.2 Overview

D(-)-isoascorbicacid **3**

H_2O_2, K_2CO_3, H_2O 77%

4

5

1.2 eq DIBAH, CH_2Cl_2, -78 °C; MeOH, -78 °C → RT 84%

6

1. 3.0 eq MgBr, THF, -78 °C
2. 1.2 eq TBSCl, ImH, THF/DMF (3 : 1), 0 °C

87%

7

8

1. AD-Mix-β, $MeSO_2NH_2$, $^tBuOH/H_2O$
2. chromatograph

79%

9a + **9b**

70 : 9

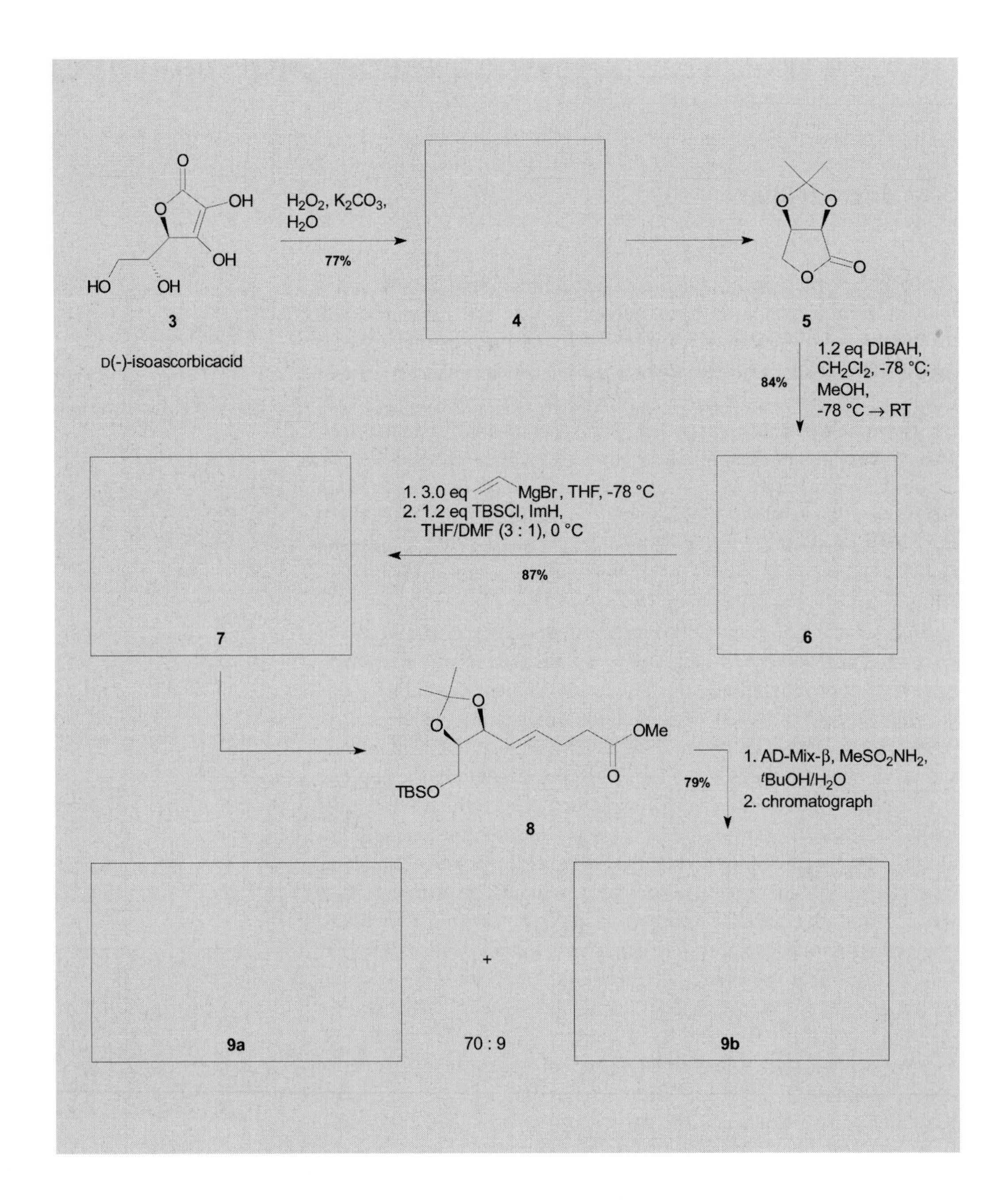

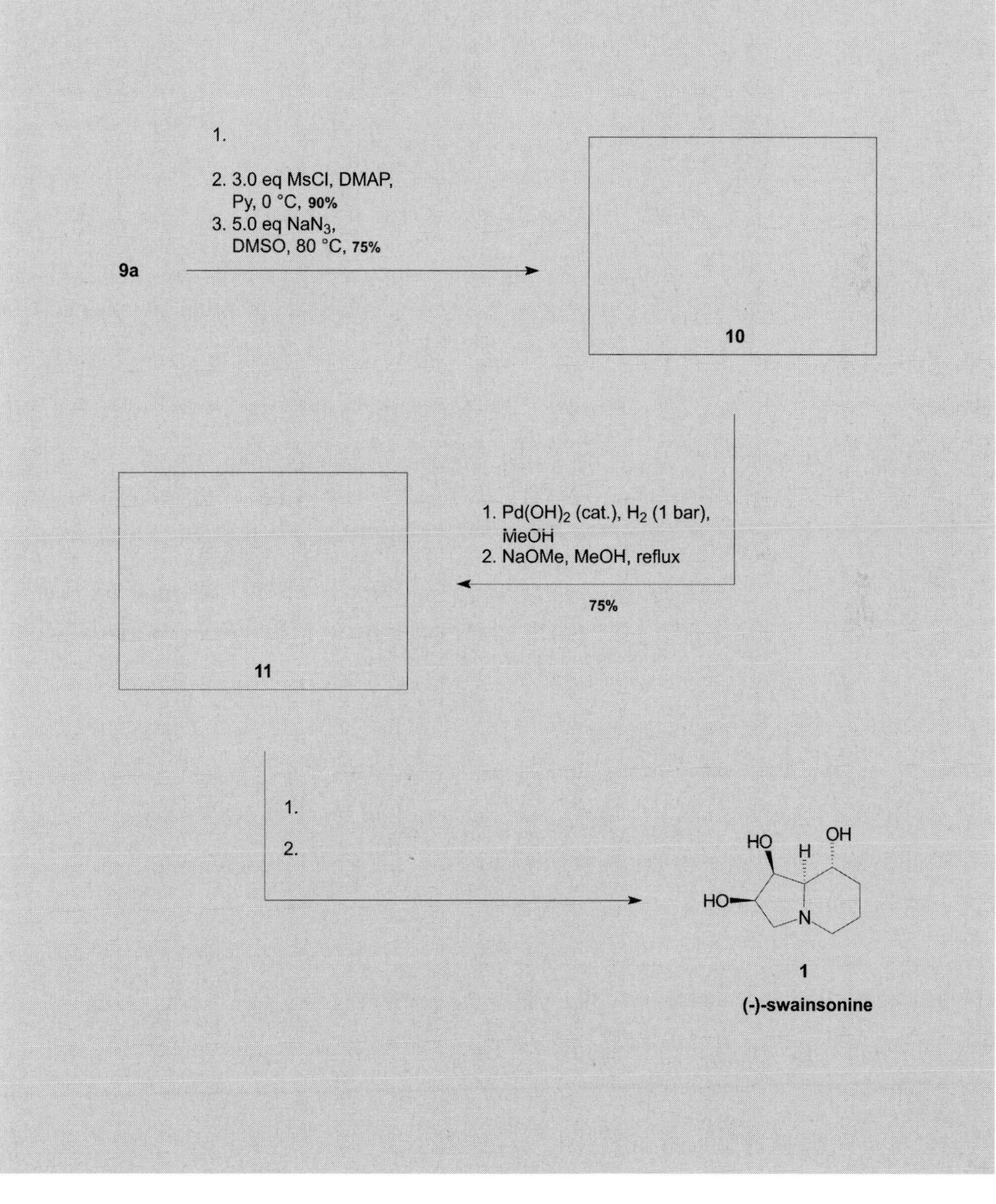

1.
2. 3.0 eq MsCl, DMAP, Py, 0 °C, 90%
3. 5.0 eq NaN3, DMSO, 80 °C, 75%
9a
10
1. Pd(OH)2 (cat.), H2 (1 bar), MeOH
2. NaOMe, MeOH, reflux
75%
11
1.
2.
HO
OH
H
HO
N
1
(-)-swainsonine

3.3 Synthesis

Problem

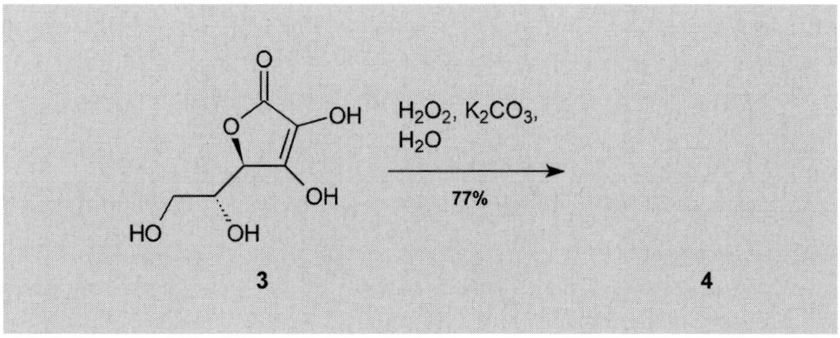

3 → 4 (H_2O_2, K_2CO_3, H_2O; 77%)

Tips

- The byproduct one obtains upon acidification is oxalic acid. Based on this, which bond must be cleaved during oxidation with alkaline hydrogen peroxide?
- Both stereogenic centers in the starting material are retained.
- The product is a lactone.

Solution

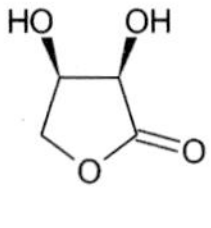

4

This oxidation, which was developed with ascorbic acid,[13] produces D-erythronolactone (**4**).[14] Hydrogen peroxide is first reduced to water and a hydroxy radical, leading to the formation from **3** of the isoascorbic acid radical **12**. The hydroxy radical oxidizes **12** to dehydroisoascorbic acid (**13**), which in a subsequent step adds hydrogen peroxide. Adduct **14** is than cleaved with base via anion **15** to compound **16**. Alkaline hydrolysis of the ester leads to hydroxy acid **17** and the byproduct oxalic acid. Subsequent lactonization produces D-erythronolactone (**4**).

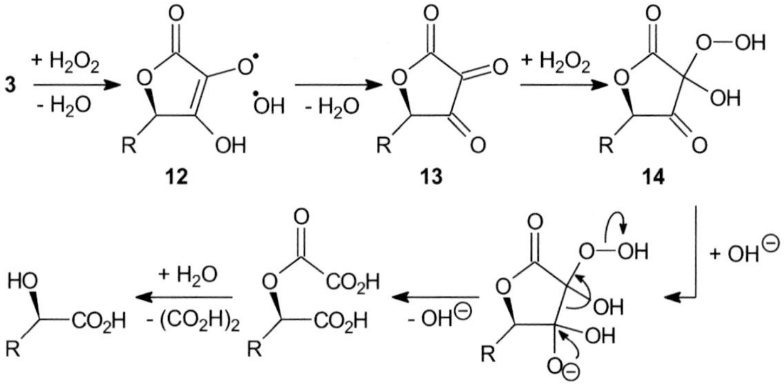

3 → 12 (+ H_2O_2, - H_2O) → 13 (- H_2O) → 14 (+ H_2O_2) → 15 (+ $OH^{\ominus}$) → 16 (- $OH^{\ominus}$) → 17 (+ H_2O, - $(CO_2H)_2$)

Problem

HO OH 4 → 5

Tip

- Which of the following sets of reaction conditions is appropriate for formation of the acetal: a) acetone, CSA; b) dimethoxypropane, PTSA, acetone; c) 2-methoxypropene, PTSA; d) 2-trimethylsilyloxypropene, HCl, CH_2Cl_2?

Solution

Dimethoxypropane (**18**) was selected for acetal formation. Transacetalization with the diol releases two equivalents of methanol, which results in a positive entropy effect.

The oldest known method for producing isopropylidene acetals is treatment of a diol with anhydrous acetone under acid catalysis. However, in order to trap the resulting water it is also necessary to include molecular sieves or copper sulfate. 2-Methoxypropene (**19**) is roughly twice as expensive as acetal **18**, but as an enol ether it is also more reactive. In especially problematic cases one can in addition resort to 2-trimethylsilyloxypropene (IPOTMS = isopropenyloxytrimethylsilane) (**20**), but for this situation it is inappropriate on the basis of cost.

MeO OMe **18**

OMe **19**

OTMS **20**

2,3-*O*-Isopropylidene-D-erythronolactone (**5**) is commercially available, but it can easily be prepared by the method described on a large scale with an overall yield of 75%, whereby diol **4** need not be purified.

Dimethoxypropane, PTSA, acetone, 75% starting from **3**.

Problem

1.2 eq DIBAH, CH_2Cl_2, -78 °C; MeOH, -78 °C → RT

84%

5 → 6

Tips

- Are esters in noncoordinating solvents reduced at low temperature by DIBAH to alcohols or aldehydes?
- The product is a bicyclic compound.

Solution

Reduction of lactone **5** results in lactol **6**, a cyclic hemiacetal. Thus, the ester function is reduced only to the level of an aldehyde.

If an aldehyde and an ester group occur together in the presence of a reducing agent like DIBAH (**22**), the aldehyde is reduced more rapidly as a consequence of its greater electrophilicity. Selective reduction of an ester to an aldehyde is therefore possible only if product **23** of the first hydride transfer does not collapse to an aldehyde. In nonpolar solvents at low temperature the tetrahedral intermediate **23** is stable and decomposes only in the course of protolytic workup. In a polar-coordinating solvent such as THF, on the other hand, the O-Al bond is weakened to such an extent by coordination of solvent with the metal atom in **24** that the aldehyde arises even before hydrolysis and is immediately reduced further to an alcohol.[15]

6 ⇌ 6a

21 22 → 23 ⇌ 24

Problem

6 → 7: 1. 3.0 eq vinylMgBr, THF, -78 °C; 2. 1.2 eq TBSCl, ImH, THF/DMF (3 : 1), 0 °C; 87%

Tips

- Lactol **6** is a hemiacetal, which is in equilibrium with its open-chain form **6a**.
- The first step is a *Grignard* reaction.
- An alcohol function is silylated in the second step.

Solution

Vinyl *Grignard* reagent reacts with the aldehyde function in the open-chain form of compound **6**, leading initially to diol **25** with a diastereoselectivity of 71:2 (*anti*:*syn*). The diasteromeric alcohols can be separated after silylation to **7**, but this is unnecessary since both isomers lead to the same product in the subsequent reaction.

The steric demand of the *tert*-butyl group in TBSCl reduces the reactivity of this silylating reagent to such an extent that not only the base imidazole but also the dipolar aprotic solvent DMF must be added. Moreover, primary alcohols react more rapidly than secondary alcohols, while tertiary alcohols are inert under these conditions.[16] It is therefore possible to distinguish between the two free hydroxy functions in diol **25**.

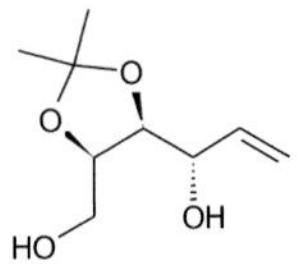

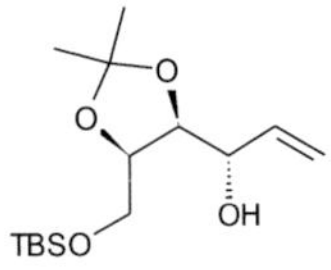

Problem

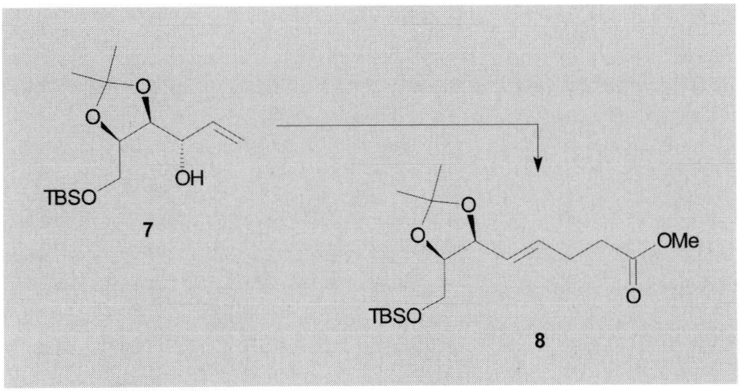

Tips

- By how many carbon atoms is the chain lengthened?
- There exists a name reaction useful for preparing γ,δ-unsaturated carboxylic acids from allylic alcohols.
- Arranging molecule **8** in a conformation in which the C–O and C–C double bonds are incorporated into a six-membered ring leads to insight into the transition state of the reaction.
- The process is a [3,3]-sigmatropic rearrangement with a chair-like transition state.
- This amounts to a variant of the *Claisen* rearrangement.
- In the *Johnson* orthoester variant a weak carboxylic acid catalyzes the first two steps of the reaction.
- Propionic acid catalyzes not only the first step, a transacetalization, but also the second, which is an elimination to a ketene acetal.

Solution

Seven equivalents of trimethyl orthoacetate and 0.3 equivalents of propionic acid are heated under reflux with allylic alcohol **7** in toluene, with concurrent distillation of the resulting methanol.

The first step in the *Johnson* orthoester method is a transacetalization of the orthoacetate leading to mixed orthoester **26**. The ketene acetal **27** required for rearrangement arises upon acid-catalyzed elimination of methanol. The necessarily six-membered transition state has a strong preference for a chair conformation (see **28**), so that reaction proceeds under kinetic control with complete *trans* selectivity relative to the newly formed double bond.

MeO, O, R

28

$H_3CC(OMe)_3$, $H^{\oplus}$(cat.), Toluene, - MeOH

7 OH TBSO O O

R O MeO OMe **26**

$H^{\oplus}$(cat.) - MeOH

R O OMe **27**

Δ

8 TBSO O O OMe

$H_3CC(OMe)_3$, propionic acid (cat.), toluene, reflux, 99%.

Problem

TBSO O O OMe O **8**

AD-Mix-β, $MeSO_2NH_2$, $^tBuOH/H_2O$, 79%

9a + **9b** 70 : 9

Tips

- AD-Mix-β is the shorthand designation for the reagent mixture employed in a very well-known method for asymmetric dihydroxylation.
- In a standard representation of the *Sharpless* dihydroxylation, the largest group is shown at the bottom left. The chiral reagent used here then attacks preferentially from above.
- The product of the dihydroxylation cyclizes spontaneously.

Solution

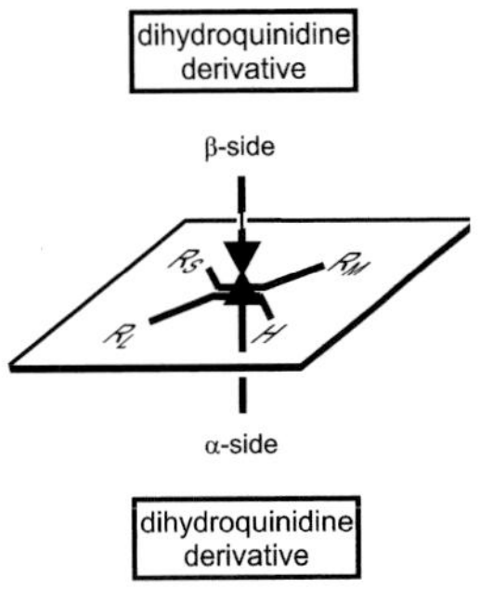

Sharpless dihydroxylation of **8** leads to diols that cyclize spontaneously to the diastereomeric lactones **9a** and **9b**, which are formed in 70% and 9% yield, respectively, and must be separated by column chromatography.

9a **9b**

Discussion

The catalytic cycle in the asymmetric *Sharpless* dihydroxylation is illustrated below.[17]

29: PHAL

30: DHQD

31: DHQ

In the currently preferred variant of this reaction, one introduces a mixture of dipotassium osmate dihydrate ($K_2OsO_4 \cdot 2H_20$) as a nonvolatile source of osmium (rather than osmium tetroxide), along with potassium hexacyanoferrate(III) ($K_3[Fe(CN)_6]$) as cooxidizing agent together with the phthalazine-(**29**)-bridged dihydroquinidine (**30**) ligand $(DHQD)_2PHAL$ or, in the case of the α-Mix, its pseudoenantiomeric dihydroquinone (**31**) analog. Potassium carbonate serves as a base. The accelerating effect of an equimolar addition of methane sulfonamide ($MeSO_2NH_2$) on hydrolysis of the osmium(VI) glycolate **33** is known as the sulfonamide effect.

The oxidation is conducted in a two-phase system with *tert*-butyl alcohol as the organic phase. The advantage of this procedure is that osmium tetroxide (**37**) is the only oxidizing agent that enters the organic phase.

Problem

1\.

2\. 3.0 eq MsCl, DMAP, Py, 0 °C, 90%

3\. 5.0 eq NaN_3, DMSO, 80 °C, 75%

TBSO, OH, **9a** → **10**

Tips

- The first reaction is a protecting-group operation.
- The strength of the Si-F bond amounts to 142 Kcal/mol, that of the Si-O bond only 112 kcal/mol.
- Is a distinction between primary and secondary alcohol functions to be expected in the mesylation?
- Azide reacts only at the most reactive position in the molecule.

Solution

MsO, OMs, **38**

The primary alcohol is released by cleavage of the silyl ether with TBAF. With HF in acetonitrile there would be a risk of cleaving the acetal group. Treatment with mesyl chloride produces dimesylate **38**. Despite a large excess of azide, substitution in the third step occurs only at the more reactive primary position.

TBAF, THF, H_2O, SiO_2, 0 °C, 84%.

Problem

1. $Pd(OH)_2$ (cat.), H_2 (1 bar), MeOH
2. NaOMe, MeOH, reflux

75%

10 → 11

Tips

- The azide is catalytically reduced in the first step.
- The second step generates the bicyclic skeleton of swainsonine (**1**).

Solution

Amine **39**, prepared by catalytic reduction, is separated from the catalyst by filtration through Celite, and the resulting methanolic solution is treated with sodium methoxide. Under these basic conditions the mesylate is replaced by an amine in an S_N2 reaction and the lactone is also opened with formation of the thermodynamically more stable lactam **11**.

39

Discussion

Reduction of amides is an important preparative method for the synthesis of primary amines. Reducing agents used for this purpose include lithium aluminum hydride, sodium borohydride, triphenylphosphine (*Staudinger* reduction), and thiols. In the present case it is important to consider the compatibility of the reduction system with the carboxylic and methanesulfonic acid functions. Platinum and palladium are often used for catalytic reduction.

Problem

1.
2.

11 → 1

(-)-swainsonine

Tips

- Which reaction sequence is preferable: acetal cleavage before or after reduction of the lactam to an amine?
- The first reaction is reduction of the lactam. A boron reagent was utilized, one with which no additives are here required.
- The acetonide is cleaved under acidic conditions.

Solution

In contrast to lithium aluminum hydride, sodium borohydride does not reduce amides. Another possible reagent would be DIBAH. However, in the present case four equivalents of borane-dimethyl sulfide complex was used as a 2M solution in THF. The amine was obtained in 94% yield after workup with ethanol.

Isopropylidene acetals are cleaved under acidic conditions, with the acid strength and reaction time varying considerably as a function of the substrate. In this case a 1M solution of the acetonide in THF was treated with an equal volume of 6N hydrochloric acid at room temperature. Final chromatographic purification was accomplished with an ion exchanger (Dowex® 1×8 200 OH^-).

1. $BH_3 \cdot SMe_2$, THF, 0 °C → RT; EtOH, RT, 94%.
2. 6N HCl, THF, RT, 96%.

3.4 Summary

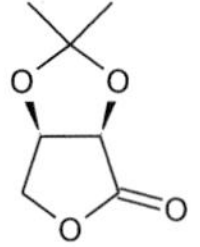

Pearson devised for his (–)-swainsonine synthesis a method that makes this pharmacologically interesting alkaloid available in 11 steps from the readily accessible lactone **5** in an overall yield of 20%. Three chromatographic purifications and five recrystallizations are required. The primary advantage over the host of other known (and in some cases very similar) total syntheses is the reproducibility of the simple transformations on a large scale. The key steps are reductive double cyclization of azido lactone **41**, which is obtained through a *Sharpless* dihydroxylation of the γ,δ-unsaturated carboxylic ester **42**. The latter is generated in a *Claisen* rearrangement (*Johnson* variant) from allylic alcohol **43**, which is in turn obtained with a vinyl *Grignard* reaction from D-erythrose **44**. The latter compound establishes the nature of two stereogenic centers in the target molecule, whereas the other two are produced in the *Sharpless* dihydroxylation.

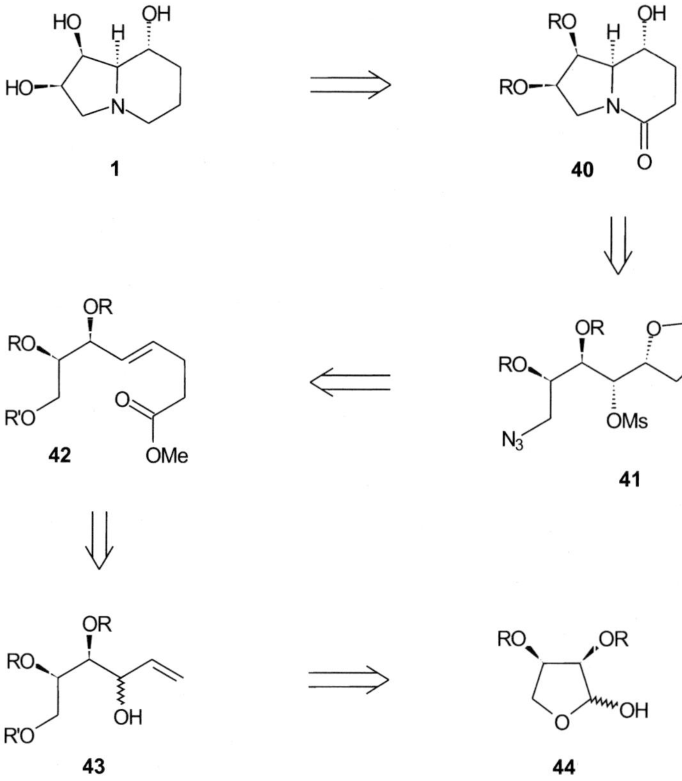

3.5 References

1 D. Davis, P. Schwaru, T. Hernandez, M. Mitchell, B. Warnock, A. D. Elbein, *Plant. Physiol.* **1984**, *76*, 972.

2 R.J. Molyneux, L.F. Fames, *Science* **1982**, *216*, 190.

3 N. Yasuda, H. Tsutsumi, T. Takaya, *Chem. Lett.* **1984**, 1201; M. Hino, O. Nakayama, Y. Tsurumi, K. Adachi, T. Shibata, H. Terano, M. Kohsaka, H. Aoki, H. Imanaka, *J. Antibiot.* **1985**, *38*, 926.

4 Übersicht über Indolizidin-Alkaloide: A.D. Elbein, R.J. Molyneux in *Alkaloids: Chemical and Biological Perspectives, Vol. 5* (Ed.: S. W. Pelletier), Wiley, New York 1987, p. 1; J. Cossy, P. Vogel in *Studies in Natural Products Chemistry, Vol. 12* (Ed.: Atta-ur-Rahman), Elsevier, Amsterdam 1993, p. 275.

5 F.P. Guengerich, S.J. DiMari, H. P. Broquist, *J. Am. Chem. Soc.* **1973**, *95*, 2055; the wrongly assigned structure has been revised (see Ref. 7).

6 S.M. Colegate, P.R. Dorling, C.R. Huxtable, *Aust. J. Chem.* **1979**, *32*, 2257.

7 M.J. Schneider, F.S. Ungemach, H.P. Broquist, T.M. Harris, *Tetrahedron* **1983**, *39*, 29.

8 To the biological effect: Y. Nishimura in *Studies in Natural Products Chemistry, Vol. 10* (Ed.: Atta-ur-Rahman), Elsevier, Amsterdam 1992, p. 495.

9 J.P. Michael, *Nat. Prod. Rep.* **1995**, 535; P.E. Goss, M.A. Baker, J.P. Carver, J.W. Dennis, *Clin. Cancer Res.* **1995**, *1*, 935; P. C. Das, J.D. Roberts, S.L. White, K. Olden, *Oncol. Res.* **1995**, *7*, 425.

10 T. Suami, K. Tadano, Y. Iimura, *Chem. Lett.* **1984**, 513; M.H. Ali, L. Hough, A.C. Richardson, *J. Chem. Soc., Chem. Commun.* **1984**, 447; G.W. J. Fleet, M.J. Gough, P.W. Smith, *Tetrahedron Lett.* **1984**, *25*, 1853; N. Yasuda, H. Tsutsumi, T. Takaya, *Chem. Lett.* **1984**, 1201.

11 W.H. Pearson, E.J. Hembre, *J. Org. Chem.* **1996**, *61*, 7217, and literature cited therein; J.A. Hunt, W.R. Roush, *J. Org. Chem.* **1997**, *62*, 1112; E.J. Hembre, W.H. Pearson, *Tetrahedron* **1997**, *53*, 11021.

12 H. Pearson, K.-C. Lin, *Tetrahedron Lett.* **1990**, *31*, 7571.

13 H.S. Isbell, H.L. Frush, *Carbohydr. Res.* **1979**, *72*, 301.

14 N. Cohen, B.L. Banner, R.J. Lopresti, F. Wong, M. Rosenberger, Y.-Y. Liu, E. Thom, A.A. Liebman, *J. Am. Chem. Soc.* **1983**, *105*, 3661; N. Cohen, B.L. Banner, A.J. Laurenzano, L. Carozza in *Organic Syntheses, Vol. 63* (Ed.: G. Saucy), Wiley, New York 1985, p. 127; J. Dunigan, L.O. Weigel, *J. Org. Chem.* **1991**, *56*, 6225.

15 R. Brückner, *Reaktionsmechanismen*, Spektrum Akademischer Verlag, Heidelberg 1996, p. 525.

16 P.J. Kocienski, *Protecting Groups*, Georg Thieme Verlag, Stuttgart 1994, p. 37.

17 R.A. Johnson, K.B. Sharpless in *Catalytic Asymmetric Synthesis* (Ed.: I. Ojima), VCH, Weinheim 1993, p. 227; H.C. Kolb, M.S. VanNieuwenhze, K.B. Sharpless, *Chem. Rev.* **1994**, *94*, 2483.

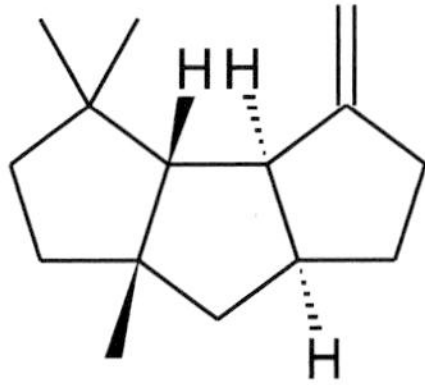

4

(–)-$\Delta^{9(12)}$-Capnellene: Shibasaki (1996)

4.1 Introduction

(–)-$\Delta^{9(12)}$-Capnellene (**1**) is produced by a soft coral known as *Capnella imbricata*, and was first isolated in 1974.[1] The compound serves the coral as protection against colonization by algae and larvae.

The capnellene (**1**) of marine origin and derivative capnellenols such as **2** belong to the family of sesquiterpenes. Similar terpenes with a tricyclic [6.3.0.0]undecane skeleton are also found in plants, including such substances as hirsutene (**3**) and corioline (**4**).[2, 3]

Capnellenes and hirsutenes are of some pharmacological interest, since they demonstrate both antitumor and antibacterial activity.

Generations of chemists have occupied themselves with the synthesis of (–)-$\Delta^{9(12)}$-capnellene (**1**). Roughly 20 total syntheses have been published covering a wide range of approaches to the compound's distinctive skeleton, which is based on three anellated cyclopentane rings. These methods include such typical cyclopentane syntheses as the *Nazarov* cyclization,[4] but also more remarkable strategies like a sequence consisting of a *Diels-Alder* reaction, [2+2]-cycloaddition, and subsequent cyclobutane ring opening.[5] Recent efforts include enantioselective radical and palladium-catalyzed zipper reactions.

The synthesis presented here, developed by *Shibasaki*'s research group, is the first total synthesis of capnellene (**1**) to take advantage of asymmetric catalysis as a way of introducing all the stereochemical information into the target compound.[8]

4.2 Overview

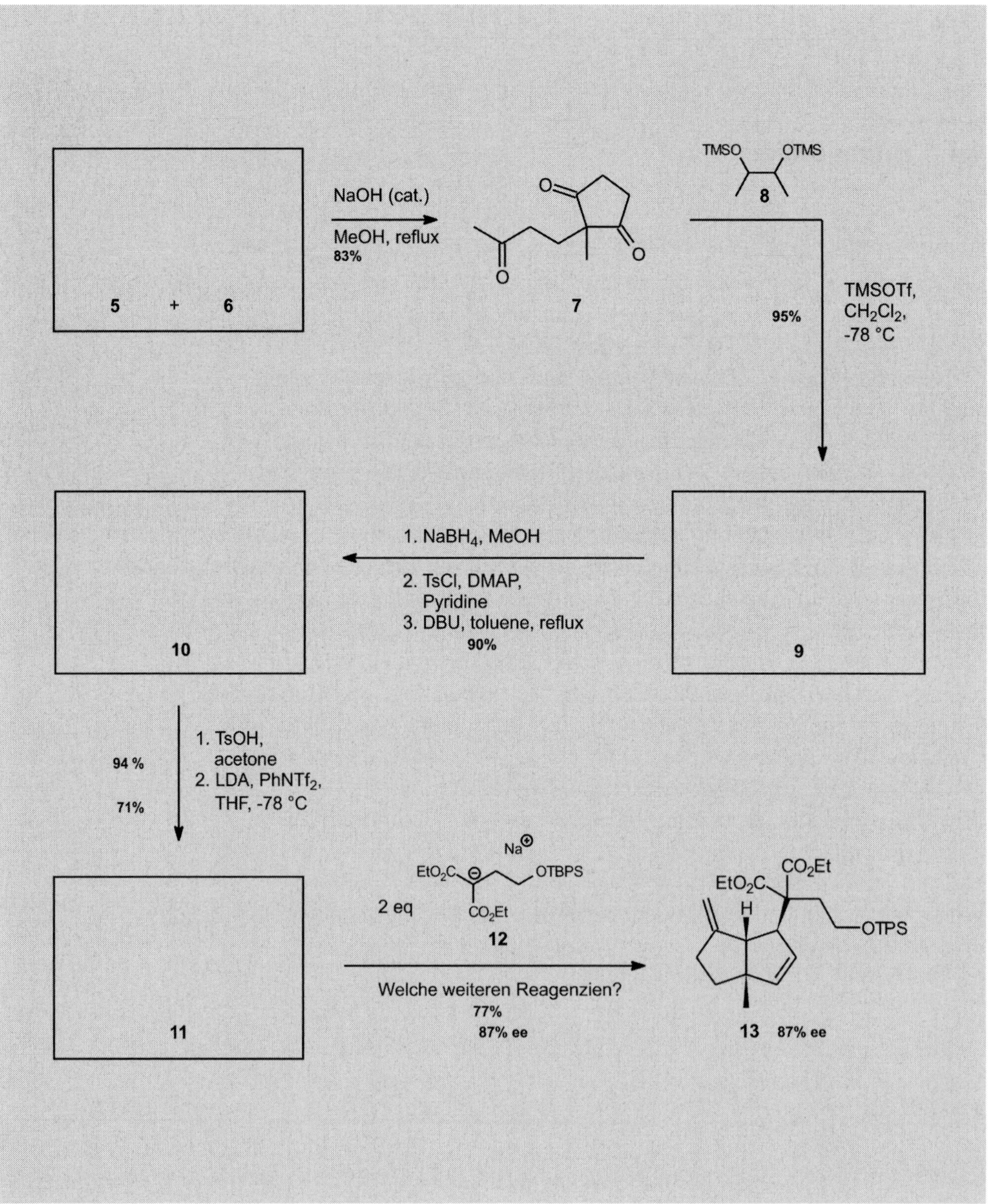
5 + 6
NaOH (cat.)
MeOH, reflux
83%
7
TMSO
OTMS
8
TMSOTf,
CH2Cl2,
-78 °C
95%
9
1. NaBH4, MeOH
2. TsCl, DMAP,
Pyridine
3. DBU, toluene, reflux
90%
10
1. TsOH,
acetone
2. LDA, PhNTf2,
THF, -78 °C
94 %
71%
11
Na⊕
EtO2C
OTBPS
CO2Et
2 eq
12
Welche weiteren Reagenzien?
77%
87% ee
CO2Et
EtO2C
H
OTPS
13 87% ee

13 → 1. 2. 3. → 14

14 → 1. 2. 3. → 15

15 → 16

16 → 1. 2. → 17

17 → 18

18 → 1. 2. → 1

1

(-)-$\Delta^{9(12)}$-capnellene

4.3 Synthesis

Problem

NaOH (cat.)

MeOH, reflux

83%

5 + 6 → 7

Tips

- Basic catalysis produces a nucleophile that participates in a C-C coupling reaction.
- The initial cyclopentane derivative is extended by a C_4 chain.
- The cyclopentane starting point is a β-dicarbonyl compound.
- Introduction of the C_4 chain is accomplished with an α,β-unsaturated carbonyl compound.

Solution

Cyclopentanedione **7** is prepared via a *Michael* addition.[9] Under basic conditions the CH acid 2-methylcyclopentanedione (**5**) adds to methyl vinyl ketone (**6**).

6 5

Problem

TMSO OTMS

8

TMSOTf, CH_2Cl_2, -78 °C

95%

7 → 9

Tips

- Reagent **8** is used for constructing cyclic systems.
- TMSOTf activates the carbonyl compound.
- The reaction itself is a protecting-group operation.
- One carbonyl group preferentially forms acetals.

Solution

The sterically less hindered carbonyl function in the side chain is selectively protected as a dioxolane. No reaction is observed with the cyclopentane carbonyl group. This procedure amounts to a modification of a method developed by *Noyori*, which permits acetal formation at low temperatures.

Discussion

The *Noyori* method is often utilized for protecting acid-sensitive substrates because, in contrast to the usual techniques for acetal formation with ethylene glycol, it does not require the presence of free acid. The original procedure employed not reagent **8**, but rather the bistrimethylsilyl ether **19** of ethylene glycol.[10] *Shibasaki* succeeded in using **19** as well on a small scale, but in larger batches the desired product **20** was subject to transacetalization to compound **21**. Introduction of the sterically more demanding reagent **8** made it possible to avoid this rearrangement.

TMSO OTMS **19**
7 → TMSOTf, CH_2Cl_2, -78 °C, 77% → **20** → **21**

Problem

9 → 1. $NaBH_4$, MeOH; 2. TsCl, DMAP, Pyridine; 3. DBU, toluene, reflux; 90% → **10**

Tips

- The first reaction is a reduction.
- Reduction of a single carbonyl group requires that only 0.25 equivalent of sodium borohydride be introduced.
- Two hydroxy functions undergo reaction in the second step.
- In the third step a bulky base attacks the newly introduced functionality from the second step.

Solution

The first step accomplishes reduction of both carbonyl groups to hydroxy functions. Only 0.25 equivalent of sodium borohydride suffices to reduce each carbonyl group. The *p*-toluenesulfonyl group introduced in the second step is a good leaving group, which is eliminated in the third step by the bulky base DBU with formation of product **10**.

Problem

1. TsOH, Acetone, 94 %

2. LDA, $PhNTf_2$, THF, -78 °C, 71%

10 → **11**

Tips

- The reaction sequence accomplishes formation of a double bond.
- The first step is a protecting-group operation.
- In the second step the strong base LDA reacts with the functionality introduced in the first step.
- Which of the two enolates is formed under these reaction conditions, the kinetically favored or the thermodynamically favored one?
- In similar reactions, trifluoromethanesulfonic anhydride is used in place of *N*-phenyl bistrifluoromethanesulfonimide.[11]

Solution

OTf **11**

In the first step the acetal of compound **10** is cleaved under acid catalysis. Kinetic control (–78 °C) in the second reaction leads to selective formation of the sterically less hindered enolate,[12] which is trapped as the enol triflate **11**.

Discussion

N-Phenyl bistrifluoromethanesulfonimide is a stable solid, and thus an easy reagent with which to work.[13] It is also well suited to the introduction of triflate groups on phenols and amines. The compound reacts with secondary aliphatic amines, but not with secondary aromatic amines.[14]

Problem

2 eq EtO_2C ⊖ OTPS, $Na^{\oplus}$, CO_2Et **12**; **11** OTf → Which further reagents? 77%, 87% ee → **13** (CO_2Et, EtO_2C, H, OTPS)

Tips

- From a mechanistic standpoint a bicyclic system is constructed first.
- Malonate **12** reacts only after the ring closure.
- The reaction is a catalytic process involving palladium.
- An allyl-Pd intermediate is trapped by malonate **12** in a nucleophilic substitution.
- Product **13**, containing three stereogenic centers, is obtained with 87% *ee*. Stereochemical information is introduced into the reaction by a chiral catalyst.

Solution

A palladium-catalyzed C-C coupling reaction – the *Heck* reaction – is used in the construction of bicyclic system **13**. Cyclization leads to a η^3-allyl-Pd complex, which undergoes nucleophilic attack by malonic ester anion **12**. This in turn leads to formation of the C_4 side chain. The mechanism of this reaction therefore differs from that of a normal *Heck* reaction.

In an oxidative addition, Pd(0) complex **22** with BINAP as a ligand accepts alkenyl triflate **11**. The resulting Pd complex **23** is cationic, since the triflate anion is bound only loosely to the palladium and dissociates from the complex.[15] *Syn* insertion of one of the two enantiotopic double bonds of the cyclopentadiene into the alkenyl-Pd bond of complex **23** leads first to η^1-allyl-Pd complex **24**. This is in rapid equilibrium with η^3-allyl-Pd complex **25**. Neither **24** nor **25** contains a β-H atom in a *syn* relationship to palladium. Moreover, internal rotation is impossible in the conformationally fixed ring system. For this reason there is no possibility of a subsequent β-hydride elimination that would once again release the palladium catalyst. In a normal *Heck* reaction (see discussion) the catalytic cycle would be broken at this point.

However, cationic allyl-Pd complexes are subject to attack by a nucleophile.[16] Such an attack releases the catalytically active Pd(0)

OTf, Me **11**; Pd, PPh_2, PPh_2 **22**; $OTf^{\ominus}$, Ph_2P, PPh_2, $Pd^{\oplus}$, Me **23**; H, $Pd^{\oplus}L_n$, Me **24**; H, $Pd^{\oplus}L_n$, Me **25** species **22**, which in turn facilitates a catalytic cycle. Malonic ester anion **12** approaches η^3-allyl-Pd complex **25** from the side away from palladium. The nucleophile is thereby introduced regioselectively at the less substituted end of the η^3-allyl-Pd intermediate, resulting in compound **13**.

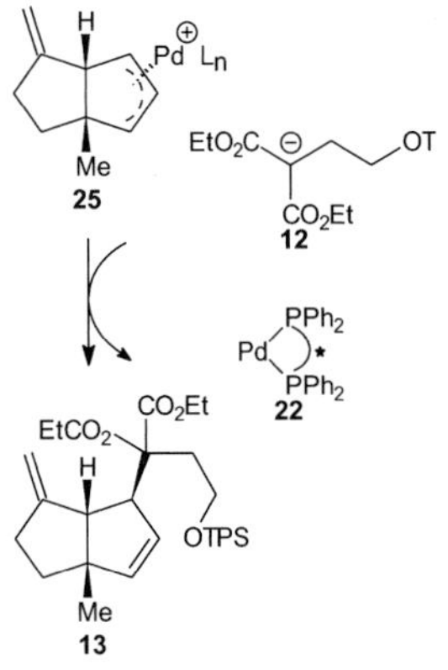

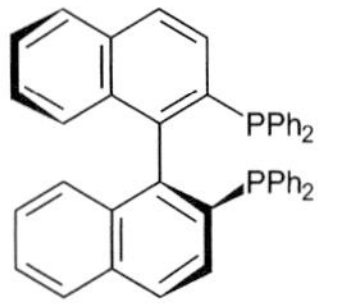

Starting material **11** for the *Heck* reaction is prochiral. The three newly established stereogenic centers in product **13** are unambiguously established relative to one another through the *Heck* and nucleophilic capture mechanisms. In the absence of further chiral information the absolute configuration of the stereogenic centers would remain undefined, and the two enantiomers **13** and *ent*-**13** would be expected to arise in equal amounts.

In order to obtain an excess of one of the two enantiomers **13** or *ent*-**13**, chiral information must be introduced into the *Heck* reaction. Axially chiral ligands at a metal atom (e.g., BINAP) are often used as bearers of chiral information in transition-metal catalyzed reactions.[17] *Shibasaki* in this case used the (*S*)-enantiomer of BINAP (**26**).

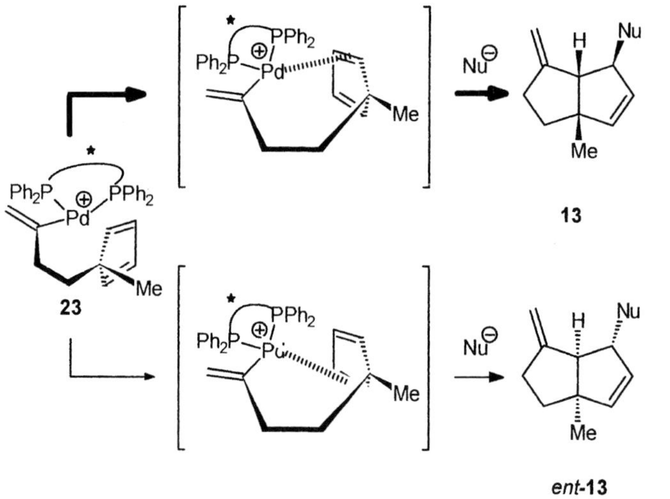

Two double bonds are available to palladium for oxidative addition in alkyl-Pd complex **23**. Both double bonds are prochiral in the absence of chiral ligands on the palladium, but they differ in their topicity. Introduction of a BINAP ligand provides a chiral environment around palladium in complex **23**, so one of the double bonds in the cyclopentadiene is attacked preferentially. This differentiation within the previously prochiral cyclopentadiene unit of compound **23** leads to an enantiomeric excess of 87% in favor of the (+)-(4*S*,10*S*,11*S*) enantiomer **13**.

Shibasaki observed a salt effect in this synthesis. Addition of two equivalents of sodium bromide increased the enantiomeric excess from 66% to 87%.
2.5 mol% $Pd(allyl)_2Cl_2$, 6.3 mol% (*S*)-BINAP, 2 eq.NaBr, 2 eq. sodium enolate **12** of the malonic ester, DMSO, RT, 77%, 87% *ee*.

Discussion

The *Heck* reaction is a palladium-catalyzed C-C bond-forming procedure that joins benzylic, vinylic, and aryl halides or the corresponding triflates with alkenes or alkynes. The result is an alkenyl- or aryl-substituted alkene. The mechanism below is assumed to apply to the *Heck* reaction.[18]

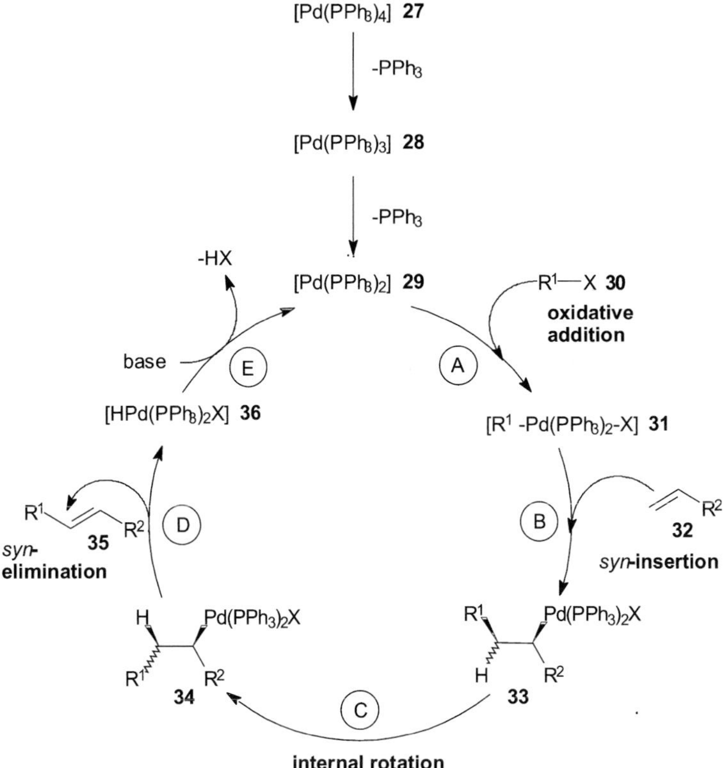

Pd(II) compounds such as $Pd(OAc)_2$ or $Pd(PPh_3)_2Cl_2$ can be used as the palladium species, and these are reduced to the corresponding catalytically active Pd(0) complexes with phosphines, olefins, or bases; alternatively, Pd(0) complexes such as $Pd(PPh_3)_4$ (**27**) or $Pd_2(dba)_3$ can be introduced directly. The catalytically effective species is the coordinatively unsaturated 14-electron complex **29**, which arises in situ from the corresponding Pd(0) compound **28** through endergonic loss of a ligand.

The catalytic cycle consists of the following basic reactions. In the first step (A) of the cycle, Pd species **29** undergoes oxidative addition of an alkenyl or aryl halide **30**. The result is a σ-alkenyl or σ-aryl palladium complex **31**. In the second step, alkene **32** coordinates with Pd(II) compound **31**. This is followed by *syn* insertion (B) of the double bond into the alkenyl or aryl palladium bond.

A β-hydride elimination requires that there be a coplanar relationship between the appropriate hydrogen atom and palladium. Since this *syn* arrangement is not present in conformation **33**, internal rotation (C) of the alkyl palladium species is necessary. β-Hydride transfer to palladium from **34** leads to *syn* elimination with formation of olefin **35**. The catalyst is subsequently regenerated through the presence of base with elimination of HX (E), after which the catalytic cycle can be repeated.

Problem

EtO_2C, CO_2Et, H, OTPS — **13** —— 1. 2. 3. ——> OBz, H, OTPS — **14**

Tips

- In the course of the reaction sequence a carboxyl group is removed.
- The method used for decarboxylation does not require prior hydrolysis.
- The decarboxylating reagent is lithium chloride.
- In the second step an ester is reduced.
- The third step is a protecting-group operation.

Solution

First, one of the two ethyl ester groups is decarboxylated by the method of *Krapcho*, in which compound **13** is refluxed with lithium chloride and water in DMSO (see Chapter 1).[19] This is followed in a second step by reduction of the remaining ester function with DIBAH to a hydroxy group,[20] which is protected in the next reaction with benzoyl chloride.

1. 2 eq. LiCl, 1 eq. H_2O, DMSO, reflux, 84%.

2. 3 eq. DIBAH, CH_2Cl_2, –78 °C, 96%.
3. 2 eq. BzCl, 0.1 eq. DMAP, pyridine, CH_2Cl_2, 0 °C, 100%.

Problem

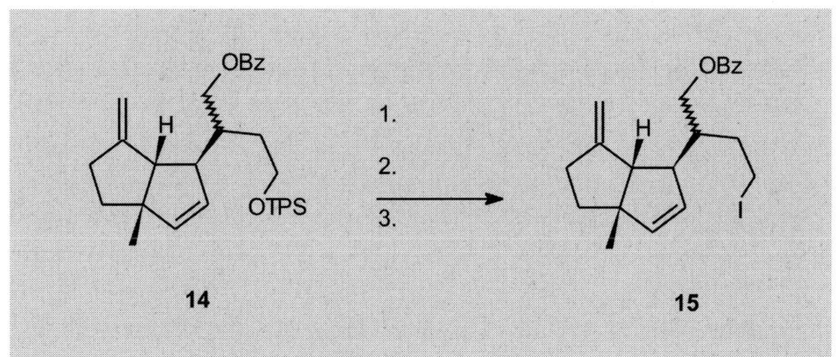

Tips

- This reaction sequence accomplishes in two steps the conversion of an alcohol into an iodide.
- In the first of the two reaction steps the protecting group is removed from the alcohol function, and the latter is then transformed into a good leaving group.
- The third step is a *Finkelstein* reaction.

Solution

The TPS group is a silyl protecting group that can be removed using TBAF as a source of fluoride ion. After removing the protecting group the alcohol function is converted into a mesylate, which is transformed in the third step via a *Finkelstein* reaction into an iodide.

1. TBAF, THF, RT, 97%.
2. 10 eq. MsCl, 15 eq. NEt_3, CH_2Cl_2, –23 °C, 100%.
3. 5 eq. NaI, acetone, RT, 100%.

Discussion

Halides are often prepared in a single step from alcohols through use of the *Appel* reaction.[21] The reagents in this synthesis are triphenylphosphine and a halogen species such as tetrachloromethane, hexachloroacetone, or iodine. In place of the *Appel* reaction it is often possible to use inorganic acid chlorides, including phosphorus tribromide or thionyl chloride (see Chapter 16).

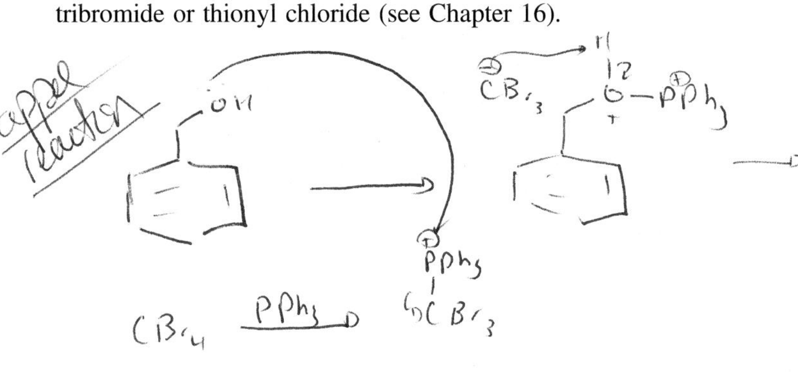

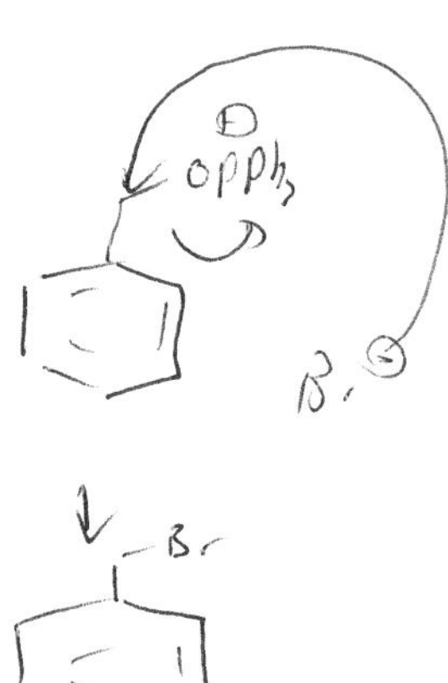

Problem

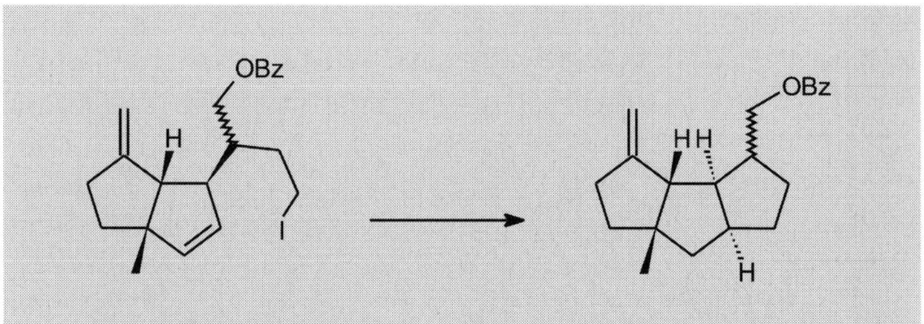

15 → **16**

Tips

- A ring closure occurs. This cannot be accomplished with the *Heck* reaction.
- The desired cyclization reaction can, like the *Heck* reaction, be carried out with haloalkenes, but it leads to saturated products.
- Organotin reagents are normally employed.
- The process occurs as a radical chain reaction.

Solution

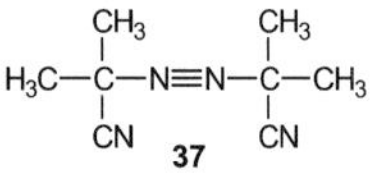

37 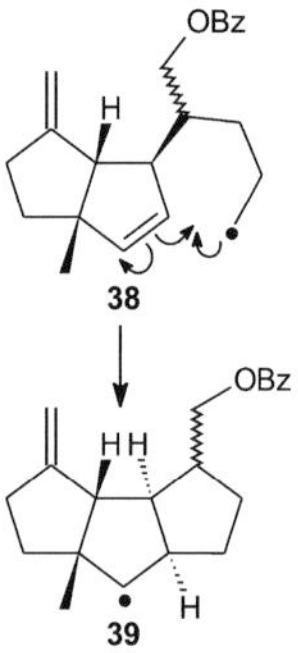 **38** **39**

A radical cyclization was conducted with 2,2′-azobisisobutyronitrile AIBN (**37**) as the radical initiator. Tributyltin hydride serves as the chain transfer reagent. Radical **38** arises from halide **15** through abstraction of an iodine atom, and this in turn cyclizes to radical **39**. Compound **39** then abstracts a proton from tributyltin hydride. The resulting tributyltin hydride radical reinitiates the radical mechanism, in that it abstracts an iodine atom from another halide molecule **15** (see Chapter 14).

Cyclization leads to formation of a new stereogenic center. However, since the radical-bearing side chain can attack the double bond only from above as a result of its β linkage with the bicyclic system, the stereochemistry of the new stereogenic center is definitively established. This new stereogenic center is thus introduced with substrate control.

2 eq. AIBN, $(^{n}Bu)_3SnH$, benzene, reflux.

Discussion

Construction of tricyclic system **16** via a *Heck* reaction is impossible for two reasons. On one hand, use of an alkyl halide with a β-hydrogen atom is excluded, because after oxidative addition the palladium would undergo immediate *syn* elimination with formation of an olefin. Therefore, *syn* insertion into the olefin component is ruled out. On the other hand, during β-hydride elimination in the normal catalytic cycle a hydride in the β position must be transferred to palladium, but there are no accessible β-hydrogen atoms present in the olefinic component of compound **15**.

Problem

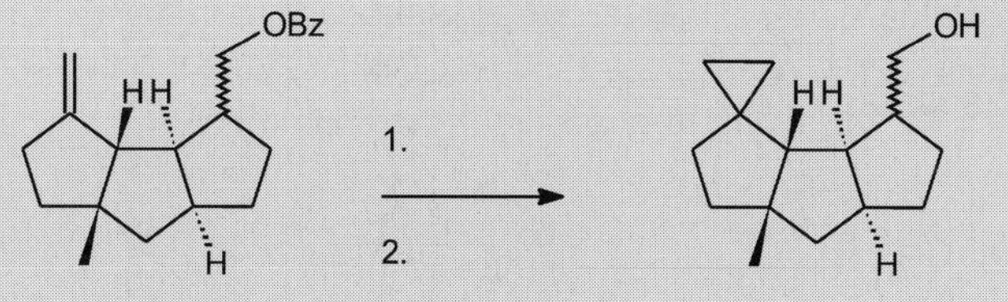

16 → 1. 2. → **17**

Tips

- In one of the two reactions a double bond is cyclopropanated. The cyclopropanation reagent in the process transfers a methylene group.
- The cyclopropanation requires a methylene carbene that can add to the double bond. Under which of the following sets of conditions can methylene be prepared in situ from diiodomethane: a) CH_2I_2, Cu/Zn; b) CH_2I_2, $KMnO_4$; c) CH_2I_2, $ZnEt_2$?
- An organozinc compound is used in the cyclopropanation.
- Certain organometallic reactions can be accelerated by coordination of the metal with a hydroxy group in the substrate. This has an influence on the preferred sequence of the desired reactions.
- The first reaction is the deprotection of an alcohol.

Solution

The first reaction is deprotection of the alcohol. Benzoate **16** is hydrolyzed with NaOH. What follows is a modified *Simmons-Smith* cyclopropanation with diiodomethane (**41**) and diethylzinc (**40**).

Et–Zn–Et (**40**) + I–CH_2–I (**41**) ⇌ Et–Zn–CH_2–I (**42**) + Et–I (**43**)

A mixture of **40** and **41** provides the active zinc species **42** in a reaction similar to the *Schlenk* equilibrium (see Chapter 1).[22] This presumably alkylates alkene **16** in a bimolecular process via the carbenoid transition structure **44**. No free carbenes arise.

44

1. NaOH, MeOH, RT, 95% over two steps.
2. 5 eq. $ZnEt_2$, 10 eq. CH_2I_2, toluene, 60 °C, 95%.

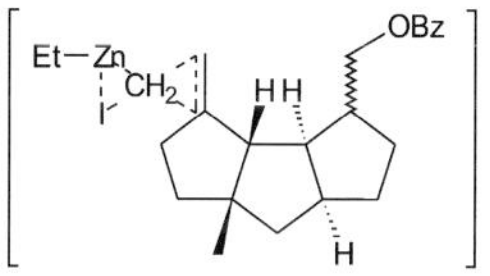

Discussion

The cyclopropanation reaction originally developed by *Simmons* and *Smith* utilized for eliminating iodine from a geminal diodide the reducing agent zinc, which had previously been activated by copper.[23]

Allylic hydroxy groups in the substrate are able not only to accelerate the *Simmons-Smith* reaction through coordination of **42**, but their active volume also allows them to direct attack of zinc species **42** selectively to one side of the double bond.[24]

Problem

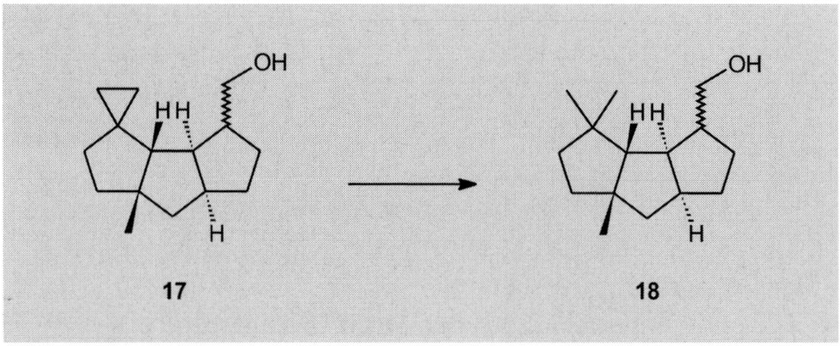

Tips

- New atoms of what element are introduced into compound **17**?
- Is this ring opening a substitution, an oxidation, or a reduction?

Solution

The sterically readily accessible external bond of the cyclopropane ring is reductively opened through hydrogenation. This can occur due to the high degree of ring strain in the three-membered ring.[25] 0.2 eq. PtO_2, H_2 (1.013 bar), AcOH, RT, 80%.

Problem

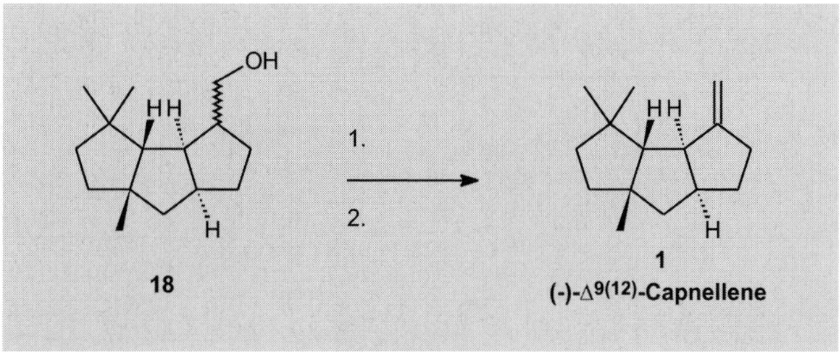

Tips

- Basic elimination of a good leaving group is not utilized in construction of the double bond.
- The second step involves a *syn* elimination similar to the *Cope* elimination of amine oxides.
- An *o*-nitrophenylselenide is first generated.
- The second reaction is oxidation to a selenoxide, which spontaneously eliminates.

Solution

The exocyclic double bonds of methylenecyclopentanes are very sensitive. Especially in the presence of acid catalysis the system readily undergoes isomerization so that the double bond assumes an *endo* position within the ring. In the creation of the *exo* methylene group *Shibasaki* utilized a mild selenoxide elimination, a reaction that occurs even at room temperature.

For this purpose the alcohol function in **18** is converted into the *o*-nitrophenylselenide **51** by treatment with *o*-nitrophenylselenocyanate (**45**) in the presence of tributylphosphine (**46**). The mechanism is assumed to be as follows:[26]

$$\underset{\mathbf{45}}{ArSeCN} + \underset{\mathbf{46}}{PBu_3} \longrightarrow \underset{\mathbf{47}}{ArSeP^{\oplus}Bu_3\ CN^{\ominus}}$$

$$\underset{\mathbf{47}}{ArSeP^{\oplus}Bu_3\ CN^{\ominus}} + \mathbf{18} \longrightarrow \underset{\mathbf{48}}{ArSe^{\ominus}} + \mathbf{49} + HCN$$

$$\underset{\mathbf{48}}{ArSe^{\ominus}} + \mathbf{49} \longrightarrow \underset{\mathbf{50}}{Bu_3PO} + \mathbf{51}$$

Oxophilic phosphonium compound **47** forms from tributylphosphine (**46**) and the phenylselenide **45**, and this then suffers nucleophilic attack by alcohol **18**. The resulting free selenium nucleophile **48** displaces phosphine oxide **50** with the formation of phenylselenide **51**. Phenylselenide **51** is oxidized by hydrogen peroxide to phenylselenoxide **52**, which at room temperature undergoes an elimination reaction to (–)-$\Delta^{9(12)}$-capnellene (**1**). The mechanism is similar to that of the *Cope* elimination, proceeding via a cyclic transition state.

1. 3 eq. 2-nitrophenylselenocyanate (**45**), 3 eq. $P(^{n}Bu)_3$, pyridine, RT.
2. H_2O_2, 10 eq. K_2CO_3, THF, RT; 78% over two steps.

45

52 → Ar–Se–OH + **1**

4.4 Summary

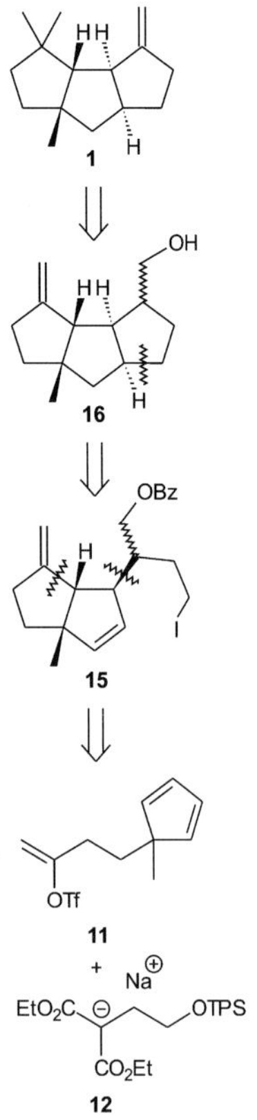

The synthesis presented here leads to (–)-$\Delta^{9(12)}$-capnellene (**1**) in 19 steps starting from triketone **7**, with an overall yield of 20% and an enantiomeric excess of 87%. *Shibasaki*'s research group employed a linear synthetic strategy. The first key step is a domino reaction consisting of a *Heck* reaction on prochiral starting material **11** followed by attack of nucleophile **12** on an allyl-Pd intermediate. This reaction leads to the construction of three stereogenic centers through asymmetric catalysis. Stereochemical information is provided by an axially chiral (*S*)-BINAP ligand on palladium.

The second key step is a radical cyclization utilized for creation of the tricyclic system **16**. The stereochemistry of the resulting newly introduced stereogenic center is established entirely by substrate control, and is thus also derived indirectly from the BINAP ligand.

Dimethyl substitution at C-1 is achieved by a *Simmons-Smith* cyclopropanation and subsequent reductive ring opening. The last step to (–)-$\Delta^{9(12)}$-capnellene (**1**) is a mild selenium oxide elimination, through which the *exo*-methylene group at C-9 is obtained.

4.5 References

1 M. Kaisin, Y.M. Sheikh, L.J. Durham, C. Djerassi, B. Tursch, D. Daloze, J.C. Braekman, D. Losman, R. Karlsson, *Tetrahedron Lett.* **1974**, *15*, 2239.

2 M. Vandewalle, P. De Clercq, *Tetrahedron* **1985**, *41*, 1767.

3 S. Takahashi, H. Naganawa, H. Iinuma, T. Takita, K. Maeda, H. Umezawa, *Tetrahedron Lett.* **1971**, *12*, 1955.

4 G.T. Crisp, W.J. Scott, J.K Stille, *J. Am. Chem. Soc.* **1984**, *106*, 7500.

5 G. Metha, A.N. Murty, D.S. Reddy, A.V. Reddy, *J. Am. Chem. Soc.* **1986**, *108*, 3443.

6 A.I. Meyers, S. Bienz, *J. Org. Chem.* **1990**, *55*, 791.

7 B.M. Trost, Y. Shi, *J. Am. Chem. Soc.* **1991**, *113*, 701.

8 T. Ohshima, K. Kagechika, M. Adachi, M. Sodeoka, M. Shibasaki, *J. Am. Chem. Soc.* **1996**, *118*, 7108.

9 C.B.C. Boyce, J.S. Whitehurst, *Helv. Chim. Acta* **1995**, *42*, 2022; Z.G. Hajos, D.R. Parrish, *J. Org. Chem.* **1974**, *39*, 1612.

10 T. Tsunoda, M. Suzuki, R. Noyori, *Tetrahedron Lett.* **1980**, *21*, 1357.

11 P.J. Stang, W. Treptow, *Synthesis* **1980**, 283.

12 J. d'Angelo, *Tetrahedron* **1976**, *32*, 2979.
13 J. E. Mc Murry, W. J. Scott, *Tetrahedron Lett.* **1983**, *24*, 979.
14 J. B. Hendrickson, R. Bergeron, *Tetrahedron Lett.* **1973**, *14*, 3839.
15 W. Cabri, I. Candiani, *Acc. Chem. Res.* **1995**, *28*, 2.
16 L. S. Hegedus, *Transition metals in the synthesis of complex organic molecules*, Wiley-VCH, Weinheim 1995.
17 R. Noyori, H. Takaya, *Acc. Chem. Res.* **1990**, *23*, 345.
18 A. de Meijere, F. E. Meyer, *Angew. Chem. Int. Ed. Engl.* **1994**, *33*, 2379.
19 A. P. Krapcho, *Synthesis* **1982**, 805.
20 E. Winterfeldt, *Synthesis* **1975**, 617.
21 R. Appel, *Angew. Chem. Int. Ed. Engl.* **1975**, 14, 801.
22 E. P. Blanchard, H. E. Simmons, *J. Am. Chem. Soc.* **1964**, *86*, 1337.
23 H. E. Simmons, R. D. Smith, *J. Am. Chem. Soc.* **1958**, *80*, 5323.
24 R. C. Gadwood, R. M. Lett, J. E. Wissinger, *J. Am. Chem. Soc.* **1986**, *108*, 6343.
25 R. W. Shortridge, R. A. Craig, K. W. Greenlee, J. M. Derfer, C. E. Boord, *J. Am. Chem. Soc* **1948**, *70*, 946.
26 P. A. Grieco, S. Gilman, M. Nishizawa, *J. Org. Chem.* **1976**, *41*, 1485.
27 W. Oppolzer, K. Bättig, T. Hudlicky, *Tetrahedron* **1981**, *37*, 4359.

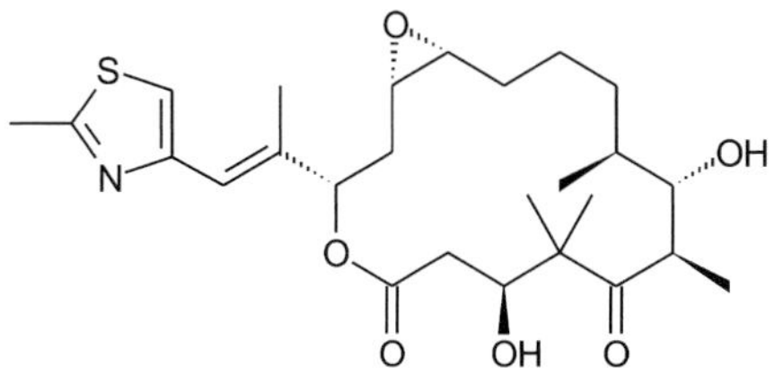

5

(–)-Epothilone A: Shinzer (1997)

5.1 Introduction

The epothilones[1] A (**1a**) and B (**1b**) were isolated and characterized by the research groups of *Höfle* and *Reichenbach* from myxobacteria of the genus *Sorganium cellulosum 90.*[2] Since discovery of the compounds in 1993 a number of total and partial syntheses have been reported. The first total synthesis was achieved by *Danishefsky*'s research group.[3] Shortly thereafter there appeared syntheses by *Nicolaou*[4] and *Schinzer.*[5]

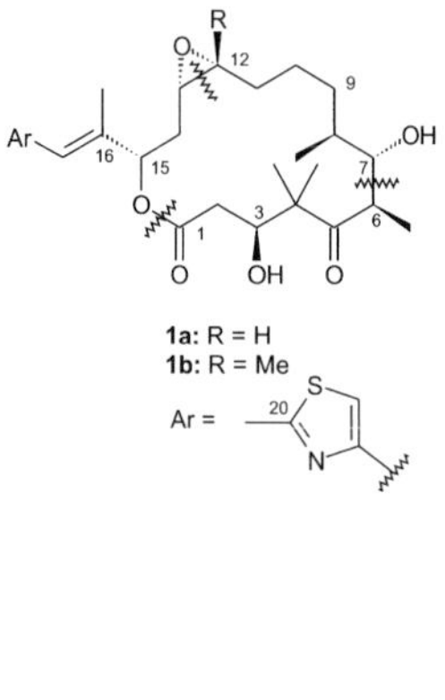

Much of the interest in the epothilones is a consequence of their fungicidal properties and especially their selective activity against tumors.[6] Their mode of action seems related to that of taxol, although the epothilone structure differs significantly from that of taxol. Similar to taxol, the epothilones bind to microtubuli[7] and thus inhibit cell division. Taxol is in fact displaced from the binding site by the epothilones. Nevertheless, the three-dimensional structures cannot be regarded as equivalent, so it is assumed that the relevant binding sites on the microtubuli are not identical but rather overlap. Epothilones display a roughly 1000- to 5000-fold higher activity relative to taxol against resistant tumor cell lines. Other advantages of the epothilones with respect to therapeutic applications are their superior water solubility and their accessibility via fermentation. Results of in vivo studies are not yet available.

1a: R = H
1b: R = Me

Total synthesis of the epothilones cannot compete with fermentation from the standpoint of quantity, but it does open possibilities for derivatization and thus the prospect of perhaps even more effective substances.[8]

Epothilone A (**1a**) can be constructed in a convergent synthesis starting from three building blocks. All the published syntheses follow essentially the same general pattern, including the *Schinzer* synthesis presented here.

5.2 Overview

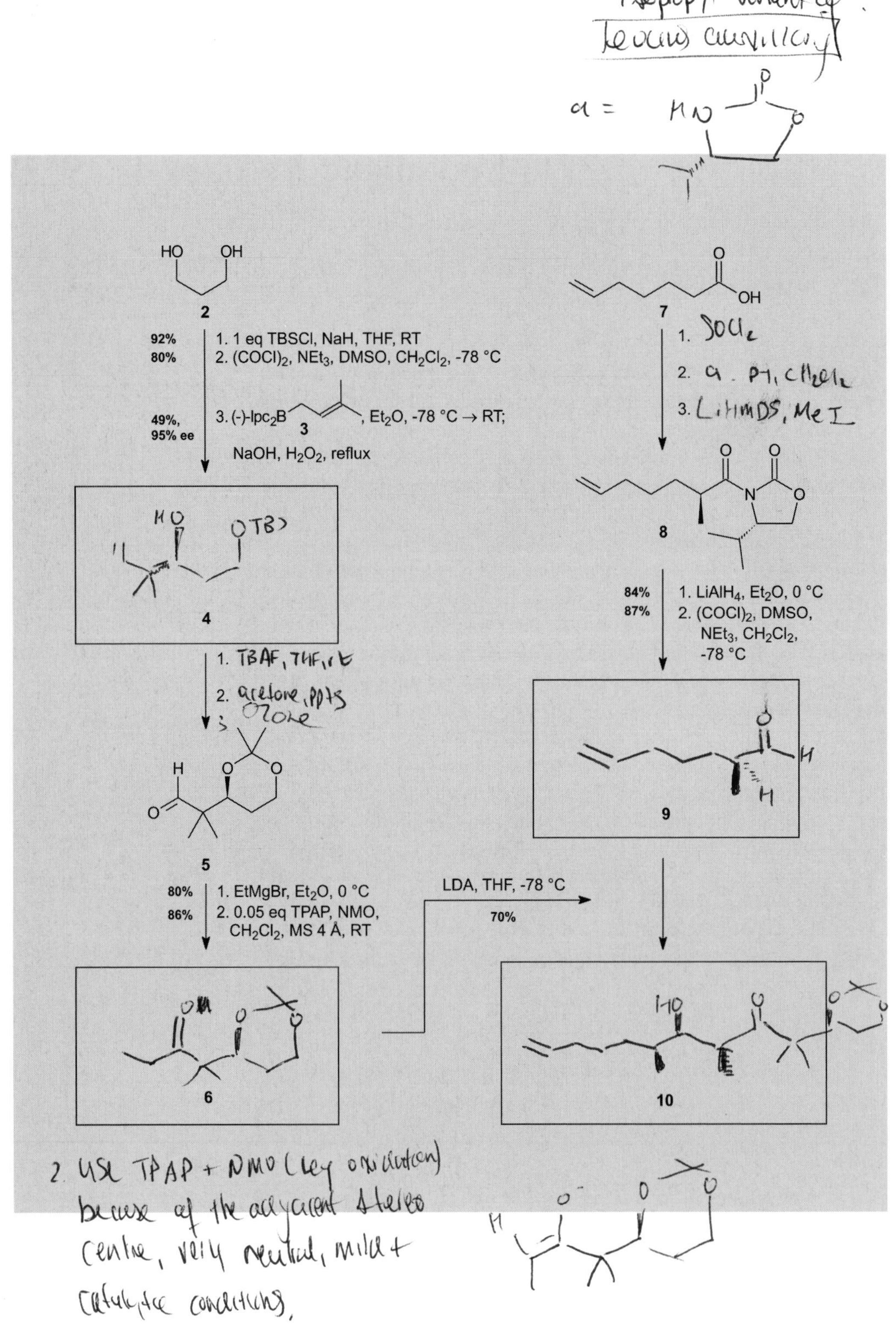

HO OH
2
92%
80%
1. 1 eq TBSCl, NaH, THF, RT
2. (COCl)2, NEt3, DMSO, CH2Cl2, -78 °C
49%,
95% ee
3. (-)-Ipc2B
3
, Et2O, -78 °C → RT;
NaOH, H2O2, reflux
4
1.
2.
H
O
O
O
5
80%
86%
1. EtMgBr, Et2O, 0 °C
2. 0.05 eq TPAP, NMO, CH2Cl2, MS 4 Å, RT
6
LDA, THF, -78 °C
70%
O
OH
7
1.
2.
3.
O
O
N
O
8
84%
87%
1. LiAlH4, Et2O, 0 °C
2. (COCl)2, DMSO, NEt3, CH2Cl2, -78 °C
9
10

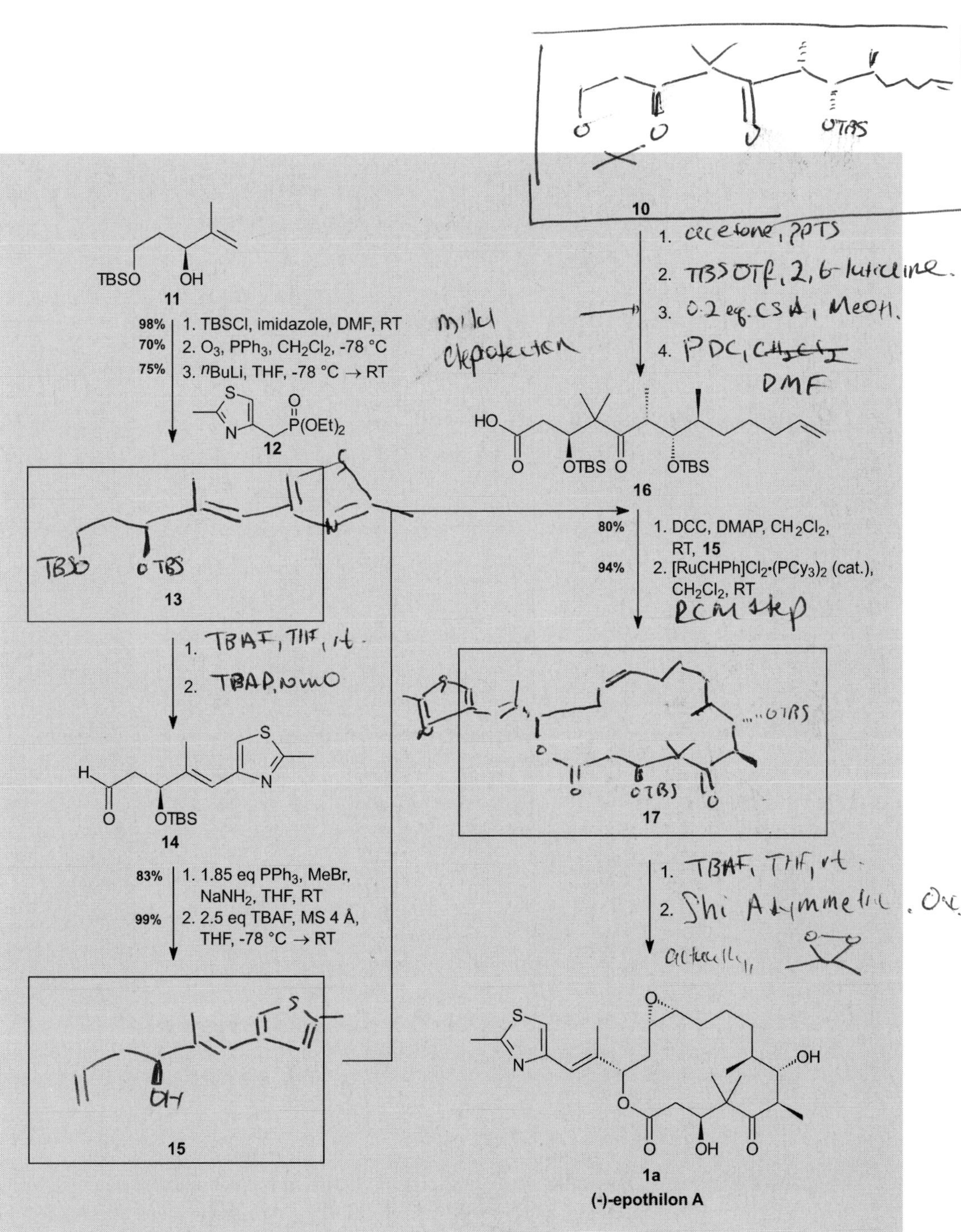
TBSO
OH
11
98%
1. TBSCl, imidazole, DMF, RT
70%
2. O_3, PPh_3, CH_2Cl_2, -78 °C
75%
3. ^{n}BuLi, THF, -78 °C → RT
P(OEt)$_2$
12
13
1. TBAF, THF, rt.
2. TBAP, NMO
H
OTBS
14
83%
1. 1.85 eq PPh_3, MeBr, $NaNH_2$, THF, RT
99%
2. 2.5 eq TBAF, MS 4 Å, THF, -78 °C → RT
15
OTBS
10
1. acetone, PPTS
2. TBSOTf, 2,6-lutidine.
3. 0.2 eq. CSA, MeOH.
4. PDC, DMF
Mild deprotection
HO
OTBS O
OTBS
16
80%
1. DCC, DMAP, CH_2Cl_2, RT, 15
94%
2. $[RuCHPh]Cl_2{\cdot}(PCy_3)_2$ (cat.), CH_2Cl_2, RT
RCM step
OTBS
17
1. TBAF, THF, rt.
2. Shi Asymmetric. Ox.
OH
1a
(-)-epothilon A

5.3 Synthesis

Problem

1. 1 eq TBSCl, NaH, THF, **92%**
2. $(COCl)_2$, NEt_3, DMSO, CH_2Cl_2, -78 °C, **80%**
3. (-)-Ipc_2B **3**, Et_2O, -78 °C → RT; NaOH, H_2O_2, reflux, **49%, 95%ee**

HO–(CH₂)₃–OH **2** → **4**

Tips

- Only one end of the diol is protected.
- Does the second step amount to an oxidation or a reduction?
- Step two is an oxidation. The crotyl reagent attacks the resulting aldehyde.
- In essence this crotylation is analogous to an aldol reaction.

Solution

O OTBS

22

Propanediol is first protected at one end as a TBS ether. The free alcohol function is then subjected to a *Swern* oxidation, leading to aldehyde **22**.

18 → $-CO_2$, CO, $Cl^{\ominus}$ → **19** → (R–CH₂–OH **2**, $-H^{\oplus}$) → **20** → ($+NEt_3$, $-Cl^{\ominus}$, $-HNEt_3^{\oplus}$) → **21** → **22** + **23**

This mild procedure utilizes DMSO as the oxidizing agent. DMSO initially reacts with oxalyl chloride, and subsequent elimination of carbon dioxide, carbon monoxide, and chloride results in the activated cation **19**. This in turn reacts with the TBS-protected alcohol to give sulfurane **20**, which eliminates a chloride ion and is thereby transformed into a sulfonium salt. The salt is deprotonated at one of the methyl groups by the base triethylamine such that finally, by elimination of dimethyl sulfide (**23**), aldehyde **22** is formed, in the course of which intramolecular migration of a proton occurs.

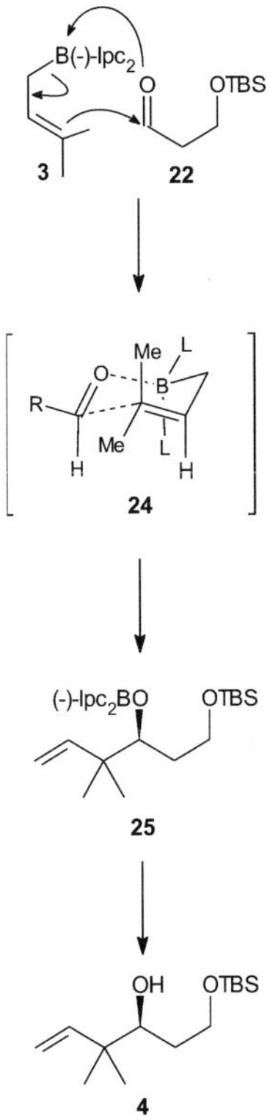

Aldehyde **22** is converted stereoselectively in the third step into homoallylic alcohol **4** using boron reagent **3**, developed by *Brown.*[10] In this case a prenyl residue is transferred.

The mechanistic course of the allylation reaction (see also Chapter 6) is analogous to that of an aldol addition. A carbonyl carbon atom undergoes nucleophilic attack by an allylic double bond, whereupon boron migrates to the carbonyl oxygen atom and a terminal double bond is created. The reaction presumably takes place through a six-membered chair-like transition state, the so-called *Zimmerman-Traxler* transition state **24**.[11] The absolute stereochemistry of the product is determined by the ligands on the boron reagent.

The resulting dialkylalkoxyborane **25** is then oxidized with alkaline peroxide and hydrolyzed to the desired alcohol **4**.

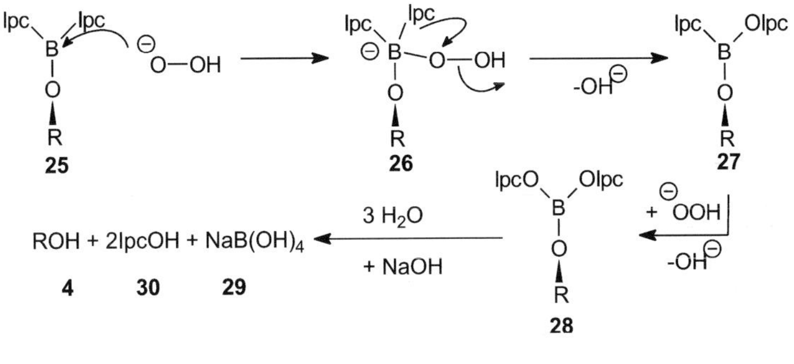

ROH + 2IpcOH + NaB(OH)$_4$

4 **30** **29**

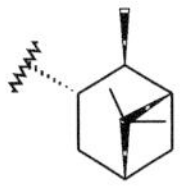

3 H_2O + NaOH

Nucleophilic attack by the hydroperoxide anion produces initially a tetrahedral borate (**26**) in which one alkyl group, here (–)-Ipc **31**, migrates from boron to oxygen, in the course of which a hydroxide anion is eliminated and a stable boron-oxygen bond develops in compound **27**. This step proceeds with retention of configuration at the migrating carbon atom. The same step are repeated with the second residue to give trialkoxyborane **28**. Finally, hydrolysis results in release of alcohol **4**, two molecules of (–)-Ipc alcohol **30**, and one equivalent of borate **29**.

Problem

Tips

- The first step effects a change in the protecting-group strategy.
- The double bond is first oxidized to a *cis* diol and then cleaved.
- The reaction is a periodate cleavage.

Solution

In a first step, the TBS ether is subjected to acid cleavage with copper sulfate in acetone containing a catalytic amount of glacial acetic acid. The resulting diol is then protected as an acetonide. Next, the double bond is oxidatively cleaved with sodium periodate and a catalytic amount of osmium tetroxide to give aldehyde **5**.

Oxidative cleavage of the olefin is accomplished by the method of *Lemieux-Johnson*.[12] The process begins with dihydroxylation of the double bond using osmium tetroxide (see Chapter 3), leading to a *cis* diol and osmium(VI) oxide. The added periodate has two functions: first, it reoxidizes the osmium(VI) species to osmium(VIII), but it also cleaves the glycol oxidatively to an aldehyde. This is the reason for utilizing several equivalents of periodate. The periodate is in turn reduced from the +VII to the +V oxidation state.

Mechanistically, what forms first is diester **32** of periodic acid. This can occur with both *cis* and *trans* diols. The ester then decomposes in a single-step process to aldehyde **5**, formaldehyde (**33**), and the trioxoiodine(V) anion **34**.

1. 2.3 eq. $CuSO_4$, HOAc, acetone, RT, 81%.
2. 0.02 eq. OsO4, 5 eq. $NaIO_4$, THF/phosphate buffer pH7, RT, 86%.

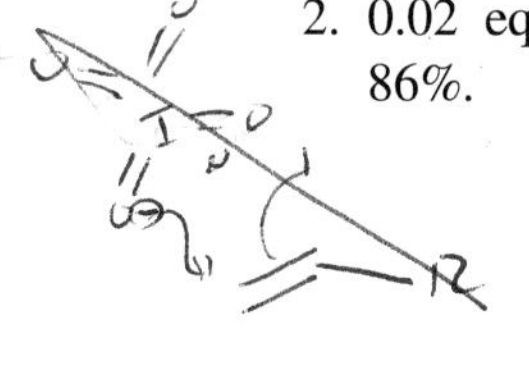

Problem

1. $EtMgBr, Et_2O$, 0 °C, 80%
2. 0.05 eq TPAP, NMO, CH_2Cl_2, MS 4 Å, RT, 86%

5 6

Tips

- The *Grignard* reagent attacks at the most electrophilic position in **5**.
- The product is an alcohol.
- Is the second step a reduction, an oxidation, or an olefination?

Solution

The *Grignard* reagent first adds nucleophilically to the carbonyl carbon atom and forms a secondary alcohol (see also Chapters 1 and 2). This is subsequently oxidized to ketone **6**. The oxidation step is based on a catalytic variant using TPAP.[13]

Similarly to oxidative olefin cleavage with periodate, the first intermediate formed is ester **35**, here a perruthenate at the oxidation level +VII. A β-elimination releases ketone **6** and the ruthenium(V) acid **36**. *N*-Methylmorpholine-*N*-oxide (NMO) serves in this case to regenerate the perruthenate(VII) species, and must therefore be introduced in stoichiometric quantity.

35 6 36

Compound **6** constitutes the first of the three building blocks.

Problem

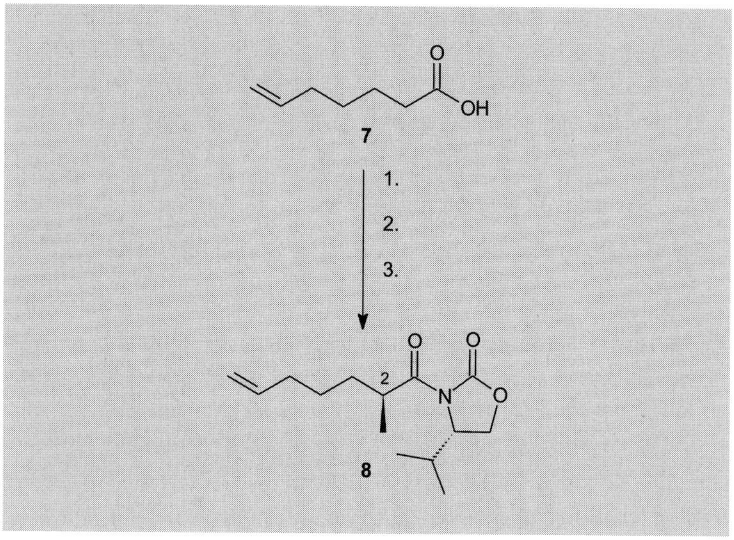

Tips

- Notice the stereochemistry at C-2 in compound **8**.
- In the first step the carboxylic acid is transformed into a reactive species.
- The oxazolidinone is acylated with the active form of the acid in the second step.
- The third step accomplishes an asymmetric alkylation.

Solution

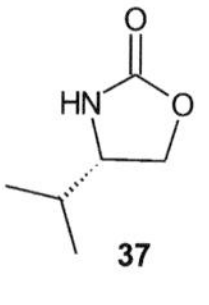

The acid is converted with thionyl chloride into the corresponding acid chloride, which reacts in a second step with the anion of the *Evans* oxazolidinone **37**[14] to give an *N*-acyloxazolidinone. This is then deprotonated with sodium hexamethyldisilazide and subsequently alkylated selectively with methyl iodide to produce compound **8**.

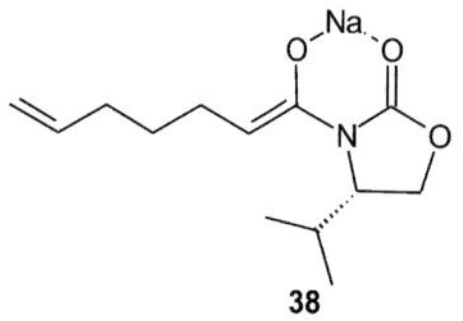

Starting with the *N*-acyl derivative, sodium hexamethyldisilazide selectively forms the *Z*-enolate **38** (amides usually form *Z*-enolates). Chelate formation between sodium and the two oxygen atoms requires a conformation for **38** in which the isopropyl group shields the bottom of the molecule, so attack by the nucleophile occurs from above. This method generally provides outstanding selectivity. The *Evans* auxiliary has also been used successfully to achieve stereocontrol in aldol[15] and *Diels-Alder* reactions.[16]

1. $SOCl_2$, benzene, reflux, 92%.
2. ^{n}BuLi, oxazolidinone **37**, THF, -78 °C, 65%.
3. NaHMDS, MeI, THF, –78 °C, 82%.

Problem

1. $LiAlH_4$, Et_2O, 0 °C, 84%
2. $(COCl)_2$, DMSO, NEt_3, CH_2Cl_2, -78 °C, 87%

8 → 9

Tips

- The *Evans* auxiliary is removed.
- In the second step *Schinzer* employed a reaction that has already appeared once in this synthesis.

Solution

The *Evans* auxiliary is first removed reductively. As a result, *N*-acyloxazolidinone **8** is transformed into a primary alcohol.

This is only one possible approach to cleavage of the auxiliary.[17] Other options include cleavage with alkaline peroxide to an acid or transamidation to the *Weinreb* amide[18] (see also Chapter 2), which could be reduced directly to an aldehyde.

Finally, the alcohol is converted into aldehyde **9** by a *Swern* oxidation.

Problem

LDA, THF, -78 °C; 9

70%

6 → 10

Tips

- The base LDA transforms **6** into an enolate.
- This enolate attacks compound **9** at the aldehyde carbon atom.
- What occurs is an aldol reaction.
- Which relative stereochemistry would you anticipate: *syn* or *anti*?

Solution

In the first step ketone **6** is deprotonated with LDA. Based on the *Ireland* model,[19] it is the *Z* enolate **39** that forms. This attacks the aldehyde carbon atom in **9** such that the *Z* lithium enolate leads to the *syn* isomer. Because of intramolecular chair-like chelate complex **39** the freedom of the transition state is limited and attack on aldehyde **9** is so directed that only the desired diastereomer **10** arises.

The observed high degree of selectivity is a result of the fact that substrate induction and reagent induction reinforce each other and are thus intensified. This is therefore a case of double stereodifferentiation.[20] The two compounds constitute what is known as a "matched pair." In a "mismatched pair" the two inductive tendencies would be in competition, and selectivity would be reduced.

Problem

Tips

- The reagent TBS chloride has already been employed previously in this synthesis under different conditions. Its role here is the same.
- The second step consists of an oxidative olefin cleavage.
- Phosphonate **12** is a compound with an acidic C-H group that is deprotonated by *n*-butyllithium.
- The third step is a *Horner-Wadsworth-Emmons* olefination.

Solution

Secondary alcohol **11** is first protected as a silyl ether with TBS chloride, after which the terminal double bond is ozonized. The resulting methyl ketone is subsequently converted stereoselectively with a *Horner-Wadsworth-Emmons* reaction[21] into olefin **13**. This reaction sequence leads to *trans* selectivity in the formation of the terminal double bond in **13**.

Ozonolysis of a double bond leads first to a so-called primary ozonide **40** through 1,3-dipolar cycloaddition. Rearrangement of primary ozonide **40** with ring cleavage produces a carbonyl oxide **42** and a carbonyl compound **41**, which then recyclize to secondary ozonide **43**. The reaction terminates with a redox process involving added triphenylphosphine, which by reducing secondary ozonide **43** releases ketone **44** and formaldehyde.

Use of a stronger reducing agent such as lithium aluminum hydride or sodium borohydride would generate the corresponding alcohols, whereas an oxidative workup with hydrogen peroxide would cause the carbonyl compounds to be oxidized in situ to carboxylic acids.

The third step is a *Horner-Wadsworth-Emmons* reaction in which a C-C double bond is created by condensation of ketone **44** with the lithium salt of the β-thiazolephosphoric acid dialkyl ester **12**. The desired compound **13** is obtained selectively with *trans* geometry. The precise mechanism of the *Horner-Wadsworth-Emmons* reaction is not yet known, so the source of the *trans* selectivity remains uncertain. It is assumed that, analogous to the *Wittig* reaction, an oxaphosphetane forms as an intermediate (see also Chapter 9).

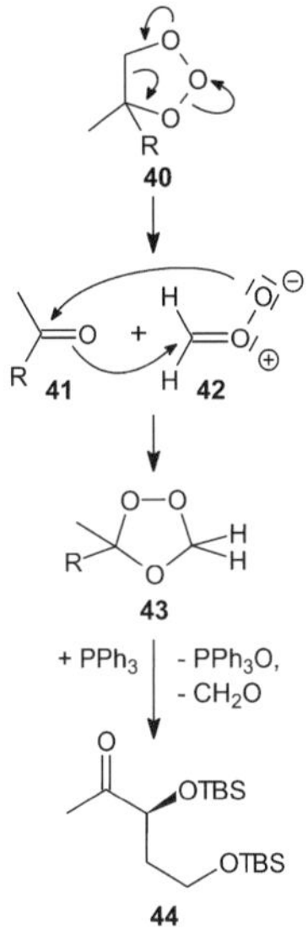

R 40 41 42 43 44 + PPh_3 - PPh_3O, - CH_2O OTBS OTBS

Problem

TBSO OTBS 13 1. 2. OTBS 14

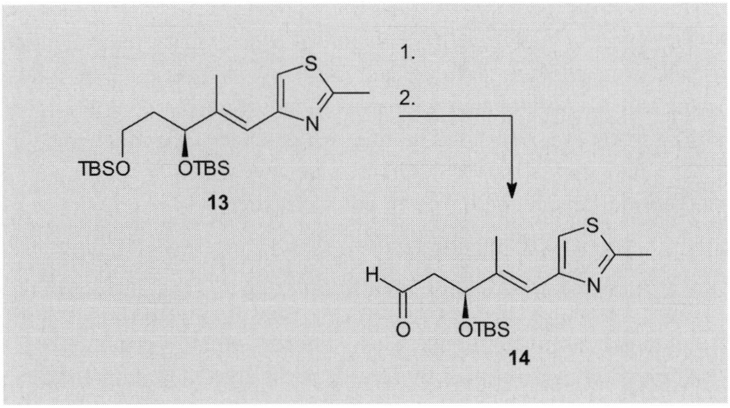

Tips

- Prior to oxidation the primary TBS ether must first be cleaved.
- This is not a *Swern* oxidation.

Solution

First, the protecting group is selectively removed from the primary (and thus more reactive) alcohol function using fluoride ion in a reaction that proceeds only in the presence of glass fragments. The fluoride in this case is introduced in the form of hydrofluoric acid. *Schinzer* has postulated in this context the participation of hexafluorosilicate as a catalyst.[22]

AcO OAc OAc

45

The newly released alcohol is then oxidized to aldehyde **14** by the mild method of *Dess* and *Martin* (see Chapter 7 for further discussion of the *Dess-Martin* oxidation).

1. HF, CH_3CN, glass fragments, 0 °C, 87%.
2. *Dess-Martin* periodinane **45**, CH_2Cl_2, RT, 78%.

Problem

1. 1.85 eq PPh_3, MeBr, $NaNH_2$, THF, RT, 83%
2. 2.5 eq TBAF, MS 4 Å, THF, -78°C → RT, 99%

14 → **15**

Tips

- The process begins with formation of a quaternary phosphonium salt from methyl bromide and triphenylphosphine, after which the salt is deprotonated with sodium amide.
- The result is a phosphorus ylide, which attacks the aldehyde.
- The first step is thus a *Wittig* reaction.
- Fluoride ion accomplishes an ether cleavage.

Solution

ylide **46** + **14** → *cis*-**47** + *trans*-**47** → (- $Ph_3P{=}O$) **48**

In the first step a *Wittig* reaction[23] is used to transform the aldehyde into a terminal olefin. This requires initial preparation of a quaternary phosphonium salt. The latter is then deprotonated with sodium amide to give phosphorus ylide **46**, which after nucleophilic attack on aldehyde **12** leads to the oxaphosphetane intermediate **47**. This intermediate in turn decomposes into olefin **48** and triphenylphosphine oxide.

Decomposition of the oxaphosphetane occurs with retention of geometry. Thus, a *cis* oxaphosphetane reacts to give a *cis* olefin and a *trans*-substituted oxaphosphetane gives a *trans* olefin. The *E*/*Z* selectivity of double bond formation plays no role whatsoever in this case, since the new linkage is an unsubstituted terminal double bond. Nevertheless, in general it is a function of the stability of the ylides. Unstable ylides usually carry alkyl residues and lead primarily to *cis* olefins. Semistable ylides typically contain aryl substituents that are able to slightly stabilize a carbanionic center, and these usually give *cis*/*trans* mixtures. Ylides bearing strongly electron-withdrawing groups (e.g., carbonyl groups) are stable and produce olefins with *trans* selectivity (see also Chapter 6). These selectivities are directly related to the ratios of the resulting oxaphosphetanes. Unstable ylides rapidly form *cis* oxaphosphetanes, but *trans* oxaphosphetanes only slowly. With semistable and stable ylides the rate of *trans*-oxaphosphetane formation increases, so the fraction of *trans* olefin also increases.

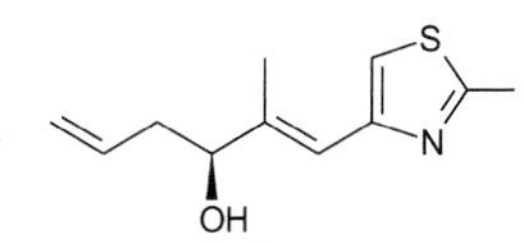

The second step is a fluoride-induced desilylation to alcohol **15**.

What follows is derivatization of **10** to **16**, which is then coupled with **15**.

OH
15

Problem

OH
10
1.
2.
3.
4.
HO
O OTBS O OTBS
16

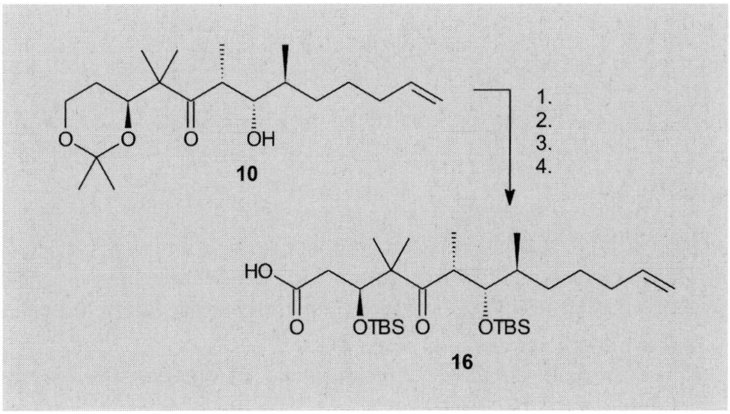

Tips

- This synthetic sequence encompasses not only protecting-group operations, but also the creation of an acid function, and this requires a particular order for the steps.
- The first step is cleavage of the acetal protecting group.
- In a second step the hydroxy functions are protected as TBS ethers.
- The third step is selective deprotection of the primary alcohol.
- Oxidation to a carboxylic acid occurs as the last step.

Solution

In the first step the cyclic acetal is subjected to acid cleavage, after which all the free alcohol functions are protected as TBS ethers with the aid of the very reactive silylation reagent TBS triflate. The primary alcohol is then deprotected under mild conditions and oxidized with PDC to a carboxylic acid.

1. PPTS, MeOH, RT, 88%.
2. 12 eq. TBSOTf, 6 eq. 2,6-lutidine (**49**), CH_2Cl_2, –78 °C, 96%.
3. 0.2 eq. CSA, MeOH, CH_2Cl_2, 0 °C, 82%.
4. 11 eq. PDC, DMF, RT, 79%.

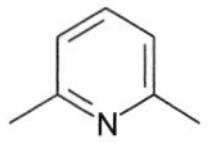

49

Problem

Tips

- An ester is synthesized in the first step.
- DCC activates the carboxyl group for esterification.
- A ruthenium complex is responsible for catalytic ring closure to a macrocyle.
- The two double bonds are transformed into one endocyclic double bond and ethene.

Solution

In the first step ester **50** is formed from secondary alcohol **15** and acid **16**. Dicyclohexylcarbodiimide (**51**) activates acid **16** for attack by the nucleophile. First the acid is deprotonated and then the resulting carboxylate **52** adds to the protonated carbodiimide species **53**, leading to an *O*-acylisourea, **54**. This activated ester is very reactive with respect to nucleophiles, and as a result of attack by alcohol **15** the urea derivative **55** is eliminated, producing the desired ester **50**.

Ring closure occurs during the next step, which takes the form of an olefin metathesis. The two terminal double bonds in **50** are coupled together with release of ethene (**58**), thereby closing the ring to **17**.

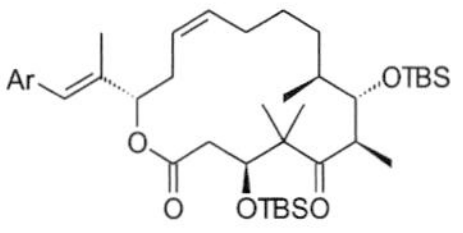

Ar, OTBS, O, O, OTBSO, **17**

Olefin metathesis[24] is an equilibrium reaction. It is assumed that in the first step olefin **50** adds to catalyst **56**. This leads via [2+2]-cycloaddition to a metallacyclobutane **57**, from which ethene (**58**) is released and equilibrium is displaced in the direction of product. Metallacarbene **59** then reacts intramolecularly in an analogous way with the second olefin to form another metallacyclobutane ring, as shown in **60**, in the decomposition of which catalyst **56** is regenerated and macrocycle **17** is released.

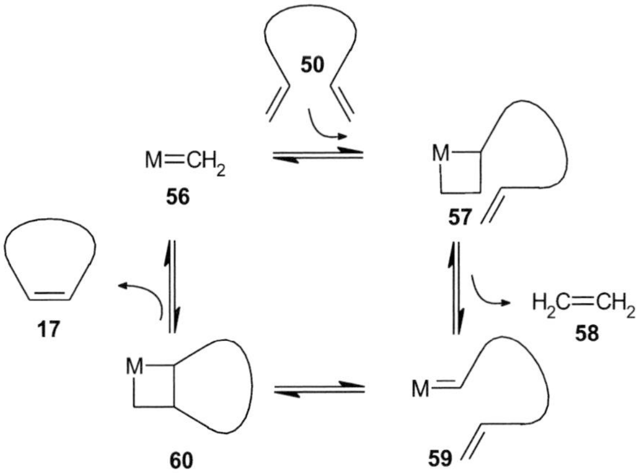

50, M=CH$_2$ **56**, M **57**, $H_2C{=}CH_2$ **58**, **17**, M **60**, M **59**

Product **17** is obtained in 94% yield, but as a 1:1 mixture with respect to the double-bond isomers.

Discussion

Two possible catalytic systems for olefin metathesis are the commercially available carbene complexes **61** and **62**. Molybdenum complex **62**, developed by *Schrock*,[25] was the first to be introduced. Nevertheless, it has the disadvantage of being both unstable and difficult to access. The newer *Grubbs* catalyst[26] **61** is now preferred because it is stable and also easy to synthesize. Both catalysts are also appropriate for starting materials containing heteroatoms (see also Chapter 16).

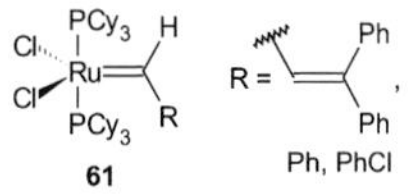

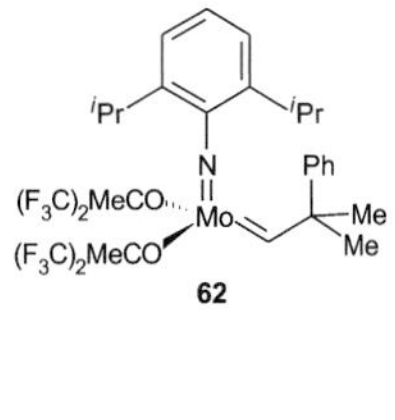

Cl, Cl, PCy_3, Ru, PCy_3, H, R, **61**; R = Ph, Ph, Ph, PhCl

^{i}Pr, ^{i}Pr, N, Ph, $(F_3C)_2MeCO$, Mo, Me, Me, $(F_3C)_2MeCO$, **62**

Problem

Tips

- The silyl protecting group is first removed.
- In a second step the macrocyclic double bond is epoxidized.
- A cyclic reagent is employed for the epoxidation.

Solution

Once the two TBS protecting groups have been removed with hydrofluoric acid as a source of fluoride, the epoxide at C-12 is constructed as the last step of the overall synthesis. For this purpose dimethyldioxirane (**63**) is utilized. It is conceivable that epoxidation could also have occurred at the exocyclic double bond at C-16, or that attack by the dimethyldioxirane could have taken place from the top face of the macrocycle. *Schinzer* obtained the desired natural product (–)-epothilone A (**1a**) in 48% yield.

1. HF, CH_3CN/Et_2O, RT, 65%.
2. Dimethyldioxirane, CH_2Cl_2, –35 °C, 48%.

5.4 Summary

(–)-Epothilone A (**1a**) was prepared with a convergent synthetic strategy starting from the three building blocks **6**, **9**, and **15**. Fragments **6** and **9** were coupled in a highly stereoselective aldol reaction (C-6/C-7), and this was followed by esterification with the last component, **15**.

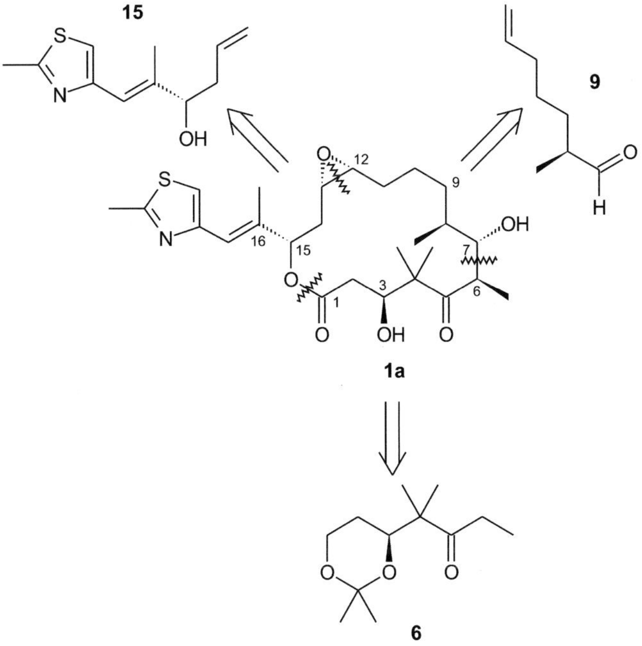

The final key step in the synthesis is a macrocyclization via olefin metathesis (C-12/C-13). The step occurs unselectively albeit in high yield and generates the double bond as an *E*/*Z* mixture with equal contributions from the two isomers.

Overall, (–)-epothilone A (**1a**) was prepared in 16 steps starting from propanediol (**2**) in a yield of 1.5%.

5.5 References

1 The name "epothilone" is derived from the terms *epo*xide, *thi*azole and ket*one*.

2 G. Höfle, N. Bedorf, H. Steinmetz, D. Schomburg, K. Gerth, H. Reichenbach, *Angew. Chem. Int. Ed. Engl.* **1996**, 35, 1567.

3 A. Balog, D. Meng, T. Kamenecka, P. Bertinato, D.-S. Su, J. Sorensen, S. J. Danishefsky, *Angew. Chem. Int. Ed. Engl.* **1996**, *35*, 2801.

4 Z. Yang, Y. He, D. Vourloumis, H. Vallberg, K.C. Nicolaou, *Angew. Chem. Int. Ed. Engl.* **1997**, *36*, 166.

5 D. Schinzer, A. Limberg, A. Bauer, O.M. Böhm, M. Cordes, *Angew. Chem. Int. Ed. Engl.* **1997**, *36*, 523.

6 IC_{50} = 2 ng mL^{-1}, where epothilone B is more active than A.

7 D.M. Bollag, P.A. McQueney, J. Zhu, O. Hensens, L. Koupal, J. Liesch, M. Goetz, E. Lazarides, C.M. Woods, *Cancer Res.* **1995**, *55*, 2325.

8 S. J. Danishefsky et al., *Angew. Chem. Int. Ed. Engl.* **1997**, *36*, 2093; K.C. Nicolaou et al., *Angew. Chem. Int. Ed. Engl.* **1997**, *36*, 2097.

9 W. Carruthers, *Some modern methods of organic synthesis*, Cambridge University Press, Cambridge 1986, p. 359.

10 U.S. Racherla, H.C. Brown, *J. Org. Chem.* **1991**, *56*, 401.

11 N. Krause, *Metallorganische Chemie*, Spektrum Akademischer Verlag, Heidelberg 1996, p. 89.

12 R. Brückner, *Reaktionsmechanismen*, Spektrum Akademischer Verlag, Heidelberg 1996, p. 505.

13 S.V. Ley, J. Norman, W.P. Griffith, S.P. Marsden, *Synthesis* **1994**, 639.

14 D.A. Evans, M.D. Ennis, D.J. Mathre, *J. Am. Chem. Soc.* **1982**, *104*, 1737; D. A. Evans, *Aldrichimica Acta* **1982**, *15*, 23.

15 D.A. Evans, J. Bartoli, T.L. Shih, *J. Am. Chem. Soc.* **1981**, *103*, 2127; M.A. Walker, C.H. Heathcock, *J. Org. Chem.* **1991**, *56*, 5747.

16 D.A. Evans, K.T. Chapman, J. Bisaha, *J. Am. Chem. Soc.* **1988**, *110*, 1238.

17 D.A. Evans, T.C. Britton, J.A. Ellman, *Tetrahedron Lett.* **1987**, *28*, 6141.

18 D.L. Clark, C.H. Heathcock, *J. Org. Chem.* **1993**, *58*, 5878.

19 R.E. Ireland, R.H. Mueller, A.K. Willard, *J. Am. Chem. Soc.* **1976**, *98*, 2868.

20 S. Masamune, W. Choy, J.S. Petersen, L.R. Sita, *Angew. Chem. Int. Ed. Engl.* **1985**, *24*, 1.

21 B.E. Maryanoff, A.B. Reitz, *Chem. Rev.* **1989**, *89*, 863.

22 D. Schinzer, Speech on the organ. chem. colloquium of the University Göttingen, 1997.

23 T. Rein, O. Reiser, *Acta Chem. Scand.* **1996**, *50*, 369.

24 H.G. Schmalz, *Angew. Chem. Int. Ed. Engl.* **1995**, *34*, 1833.

25 R.R. Schrock, J.S. Murdzek, G.C. Bazan, J. Robbins, M. DiMare, M. O'Regan, *J. Am. Chem. Soc.* **1990**, *112*, 3875.

26 S.T. Nguyen, R.H. Grubbs, J.W. Ziller, *J. Am. Chem. Soc.* **1993**, *115*, 9858.

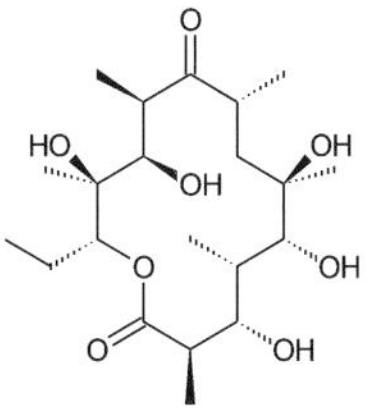

6

Erythronolid A: Hoffmann (1993)

6.1 Introduction

Total syntheses of the highly effective antibiotics erythromycin A (**2**) and (**B**) thread their way through the entire history of modern organic synthetic chemistry.[1,2] Despite their clinical significance, however, there is actually no immediate necessity for developing efficient syntheses of the compounds, since adequate amounts are available through fermentation with a strain of *Streptomyces erythreus*. Since *Corey*'s first preparation of enantiomerically pure erythronolid B (**4**) in 1978[3] and *Woodward*'s synthesis of erythromycin in 1981,[4,5,6] publications on this subject have concentrated primarily on the further development of stereoselective syntheses.[1] Attempts to synthesize derivatives of the erythromycins often have had the important goal of demonstrating the advantage or efficiency of a particular method. From a structural standpoint, **2** and **3** consist on one hand of two sugar units and on the other of a 14-membered macrocyclic lactone in each case. The lactones have been designated erythronolid A (**1**) and B (**4**), and this particular chapter is devoted to a synthesis[7,8,9] of erythronolid A (**1**) reported by *Hoffmann* in 1991.

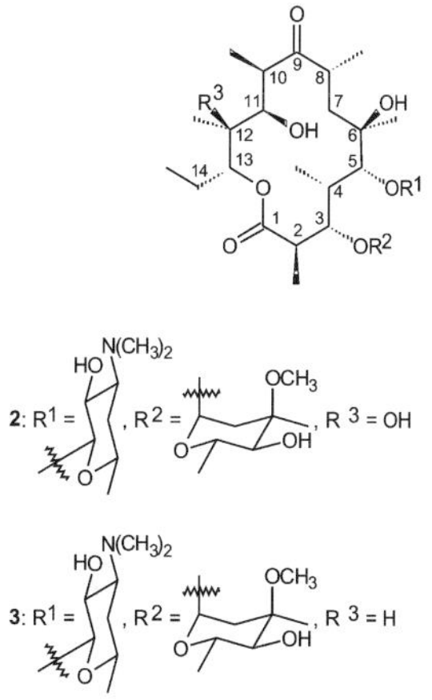

2: R^1 = , R^2 = , R^3 = OH

3: R^1 = , R^2 = , R^3 = H

1: R^1 = H, R^2 = H, R^3 = OH

4: R^1 = H, R^2 = H, R^3 = H

The first retrosynthetic scission is cleavage of the lactone into a suitably protected acid **5**. Construction of the corresponding carbon skeleton should now be possible starting with only two elementary structural elements. In the process, the stereogenic centers at C-2, C-3, C-4, C-5, C-8, C-10, and C-11 are generated by the attack of carbon nucleophiles on carbonyl groups, while the remaining centers (at C-6, C-12, and C-13) are created by epoxidation of appropriate alkenes.

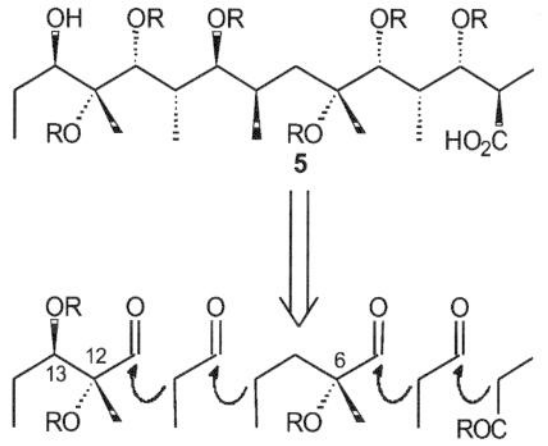

6.2 Overview

6 → 7: 1. 2.

8 → 9: 1. 2.

9 → 10: $SnCl_4$, CH_2Cl_2, -78 °C → -30 °C, 85%

10 → 11: 1. 2. 3. O_3, CH_2Cl_2, -78 °C; PPh_3

11 → 12: 1. 2.

12 → 13: 1. 2.

PMPh

CO_2Et

13 → 14

CH_2OH

14 → 15:
1. ^tBuOOH, $Ti(O^iPr)_4$, (+)-DMT, CH_2Cl_2, -30 °C — 90%, dr 8:1
2. NMO, cat. $^nPr_4N^+RuO_4^-$ — 96%
3. petroleum ether, **7**, 10 kbar — 79%

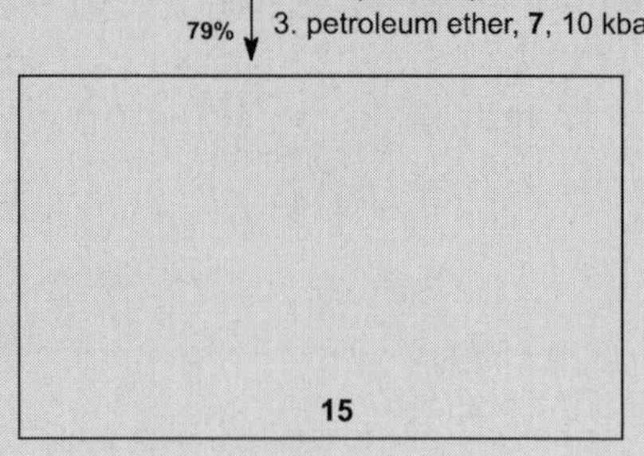

15 → 16:
1. 1.0 eq ^iPrMgCl; 5.0 eq $LiAlH_4$, THF, 65 °C, 6 h, 97%
2. PMBCl, NaH, DMF, 72%

16

15

16 → 1. 2. 3. 4. → 17

17 → 1. 2. 3. → 18

18 → 1. 2. 3. → 19

19 → 1. CSA, 0 °C, CH_2Cl_2, (4-Me-C$_6$H$_4$)CH(OMe)$_2$, **80%**; 2.; 3. → **1** **erythronolid A**

6.3 Synthesis

Problem

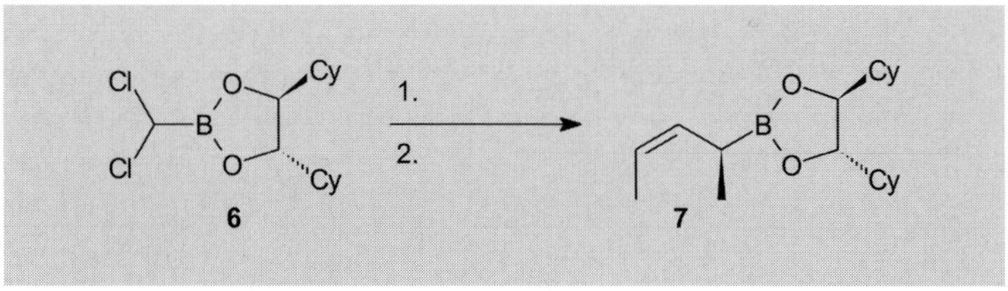

Tips

- Both reactions are nucleophilic substitutions.
- The methyl group is introduced first.

Solution

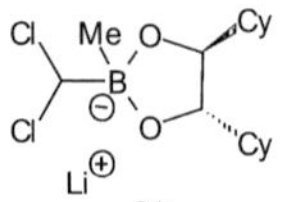

With the help of zinc dichloride, dichloride **6** is treated first with methyllithium and then with *Z*-propenyllithium. In the process, **6** reacts at –78 °C initially with methyllithium to give *at* complex **20**, which at higher temperature decomposes to the substitution product. The mechanism of the second reaction is analogous. If one interchanges the two synthetic steps what results is an intermediate (a-chlorcrotyl)boronate, which proves unstable with respect to isomerization of the double bond and epimerization of the stereogenic center in the allylic position.[13]

1. MeLi, THF, –78 °C; $ZnCl_2$, –78 °C → RT.
2. *Z*-Propenyllithium, THF, –78 °C → RT, 62%.

Problem

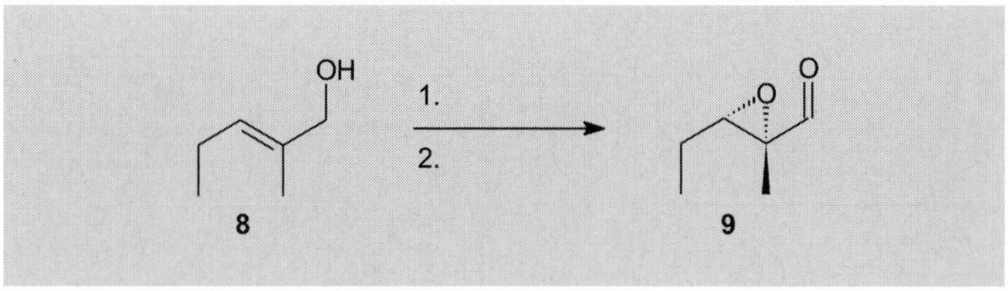

Tips

- An asymmetric epoxidation and an oxidation are carried out; in what sequence is this done?
- Oxidation of the alcohol entails use, among other things, of DMSO.

Solution

Allylic alcohol **8** is converted into epoxyaldehyde **9** with the aid of a *Sharpless* epoxidation[10, 11] followed by a *Swern* oxidation[12] (see Chapter 5). Since an allylic alcohol functionality is essential for *Sharpless* epoxidation, oxidation of the alcohol cannot be the first step. A hydroxy group in the allylic position in this case not only provides good selectivity but also increases the rate of reaction, so that allylic alcohols can be selectively epoxidized in the presence of isolated double bonds.

1. $Ti(O^iPr)_4$, L-(+)-dimethyltartrate, ^tBuOOH, MS 3 Å, CH_2Cl_2, –25 °C, 85%.
2. DMSO, $(COCl)_2$, NEt_3, CH_2Cl_2, –78 °C → 0 °C, 95%.

Discussion

The success of this approach is a result of the versatility of the *Sharpless* epoxidation of allylic alcohols,[10, 11] which also has the advantage of predictable stereochemistry.

D-(-)-tartrate
L-(+)-tartrate
21 **22** **23**

Problem

$SnCl_4$, CH_2Cl_2, -78 °C → -30 °C
85%
9 **10**

Tips

- What occurs is nucleophilic attack by the oxygen atom of cyclopentanone on the epoxide.
- An acetal is then formed.

Solution

Lewis-acid catalysis causes the ketone to attack the sterically more accessible C-3 carbon atom of epoxyaldehyde **9**, which leads to compound **10** with inversion of the configuration at this stereogenic center.[8]

9 **10**

Discussion

Compound **10** is an interesting building block for the synthesis of several pharmacologically significant macrocyclic lactones.[8] The method shown provides both enantiomeric forms of **10** enantiomerically pure in only three steps.

Problem

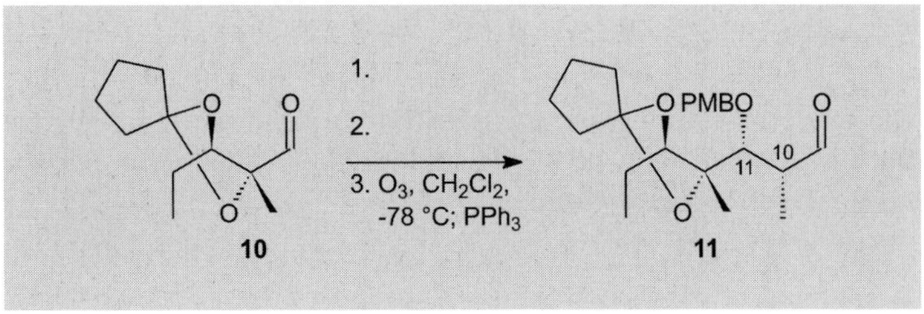

Tips

- The last step is an ozonolysis, which generates the aldehyde function from an alkene.
- What is the reaction called that introduces the olefin functionality into compound **10**? Is it a) an aldol reaction; b) an allylation; c) an alkylation; or d) a crotylation?
- The process is a crotylation.
- Through reagent control this reaction produces selectively two stereogenic centers.
- Should the reagent be a) **7**, or b) ***ent*-7**?
- Induction is determined exclusively by the allylic stereogenic center of the crotylboron reagent.
- Reaction proceeds through a 6-membered transition state.
- A *p*-methyoxybenzyl protecting group is introduced in the second step.

Solution

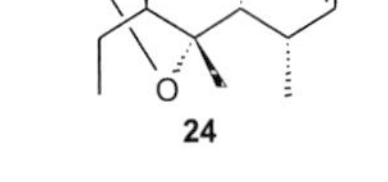

In the first step, reagent **7**, developed by *Hoffmann*, is used for asymmetric crotylation in order selectively to create the stereogenic centers at C-10 and C-11 (*dr* >98:2).[13, 14] This produces compound **24**, which is subsequently protected at C-11 as the *p*-methoxybenzyl ether. Protection is accomplished by reaction of the anion of the alcohol with commercially available *p*-methoxybenzyl chloride.[15, 16]

The allyl and crotyl boron reagents are so-called Type I reagents.[17] A cyclic six-membered transition state forms because the metallic center possesses sufficient Lewis acidity to coordinate with the oxygen atom of the aldehyde. The result is that *Z*-pentenylboron compounds give *syn* products, whereas *E*-pentenylboron compounds give *anti* products (stereodivergence). In cases where the metallic center does not show sufficient Lewis acidity (e.g., with crotyltrimethylsilane), one refers to Type II reagents. These react by way of open-chain transition states, always producing *syn* products (*syn* stereoconvergence). Type III reagents contain transition metals such as titanium and react to give *anti* products (*anti* stereoconvergence).

In the example here, the six-membered transition state **27** is destabilized by the influence of the stereogenic center through a powerful 1,2-diaxial interaction. It is worth noting that this one stereogenic center alone is demonstrably responsible for the high degree of induction.[13]

Type I: Lewis acidic, cyclic transition state stereodivergent.

Type II: not lewis acidic, open transition state, stereoconvergent *syn*.

Type III: transition metal, stereoconvergent anti.

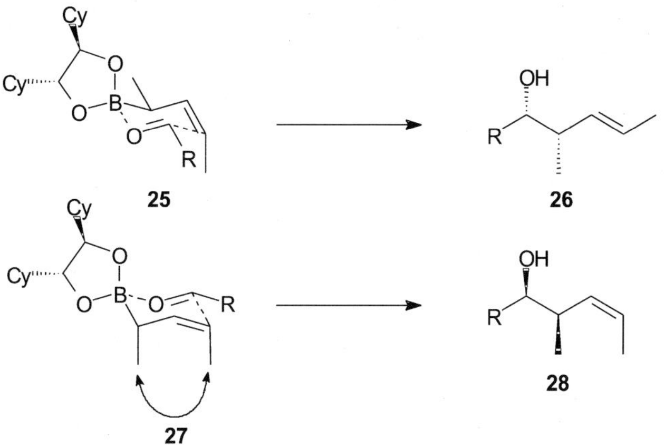

2. Benzene, **7**, 80 °C, 81%.
3. PMBCl, NaH, DMF, RT, 97%.

Discussion

The products of crotylation and aldol reactions of aldehyde **30** to give **31** and **33** are shown below. Ozonolysis makes it possible to transform **31** into **33**. This leads to the possibility of replacing the aldol reaction by a reaction sequence of crotylation and ozonolysis. With respect to allylation reactions and ozonolysis, see Chapter 5.

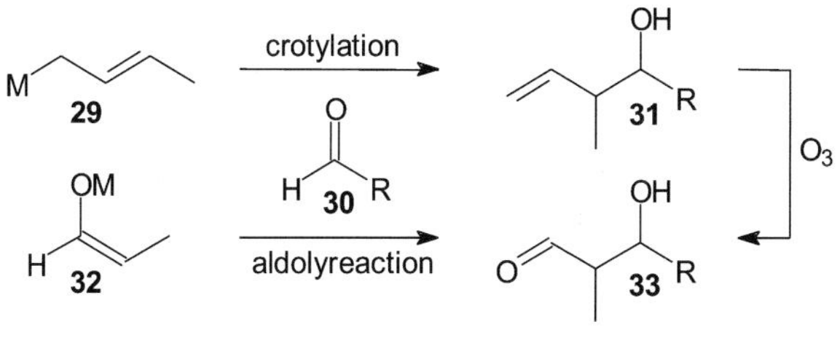

Problem

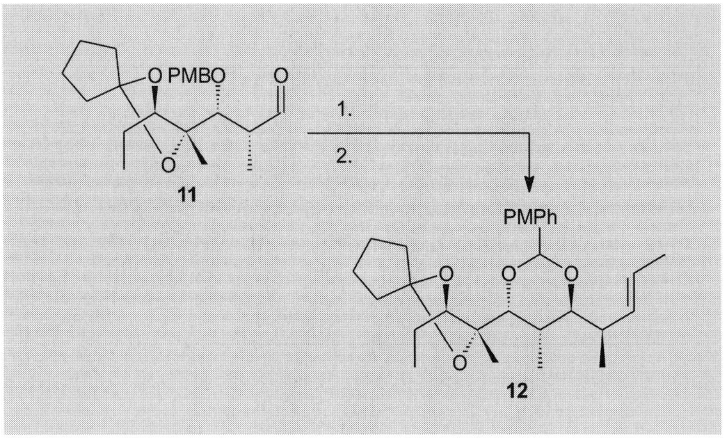

Tips

- A diol protecting group is prepared from the protecting group already present. What type of reaction does this represent: a) reduction; b) oxidation; c) nucleophilic substitution; or d) elimination?
- The method for generating the stereogenic centers has already been employed in this chapter.

Solution

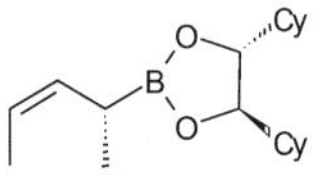

Aldehyde **11** is first crotylated with the ***ent*-7** enantiomer of the reagent discussed above, leading to compound **34**. Mild oxidation of the PMB protecting group with DDQ results in the *p*-methoxyphenyl protecting group, which corresponds to an acetal of anisaldehyde.[15, 16]

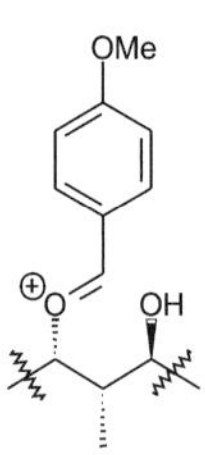

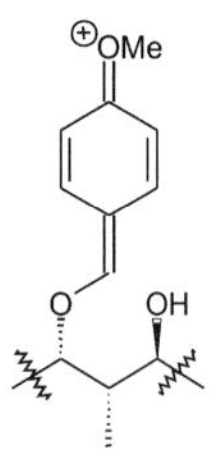

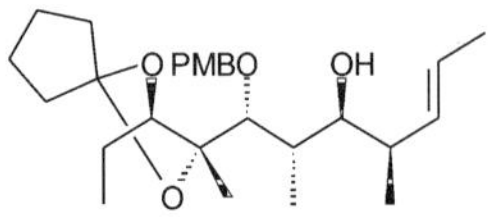

Oxidation with DDQ gives the oxocarbocation **35** pictured in the margin, which is attacked nucleophilically by the oxygen atom of the free hydroxy group.[7,18] The *p*-methoxybenzyl ether is thereby transformed selectively into a diol protecting group. This can be deprotected not only with acid (as with other acetals) but also, for example, oxidatively with DDQ or CAN. These cleavage methods can also be applied to the PMB protecting group.[15,16]

1. Petroleum ether, ***ent*-7**, MS 3 Å, 25 °C.
2. DDQ, 0 °C, MS 3 Å, 79%.

Problem

Tips

- There is an extra methyl group present at C-6.
- The bond is first cleaved between C-6 and C-7.
- A double bond must therefore be reestablished in the next step. Which of the following methods of alkene synthesis appear reasonable: a) *Wittig*; b) *Horner-Wadsworth-Emmons*; or c) *Peterson* olefination?

Solution

An ozonolysis of the double bond is carried out initially, and the resulting aldehyde is then subjected to a *Wittig* reaction. *Wittig* reactions with stable ylides usually occur with greater than 90% *E* selectivity.[19,20] The other two olefination methods can also generally be applied to the synthesis of α,β-unsaturated esters.[21,22]

1. O_3, CH_2Cl_2, –78 °C; PPh_3.
2. CH_2Cl_2, $Ph_3{}^+CH^-(CH_3)CO_2Et$, RT, 91%.

Problem

Tip

- Which reducing agent should be used to transform **13** into **14**: a) LAH or b) DIBAH?

Solution

Ester **13** is reduced with LAH to allylic alcohol **14**. Reductions of α,β-unsaturated carbonyl compounds with LAH do not always lead, as shown here, to allylic alcohols (1,2-reduction). In some cases one observes simultaneous or even more rapid reduction of the double bond (1,4-reduction). Success in this transformation can be absolutely assured by reduction with DIBAH.[23,24] In cases of doubt one should always carry out exploratory experiments.
$LiAlH_4$, Et_2O, RT, 99%.

Problem

PMPh

OH

7

6

14

90%, dr 8:1 — 1. ^tBuOOH, $Ti(O^iPr)_4$, L-(+)-dimethyltartrate, CH_2Cl_2, -30 °C

96% — 2. NMO, cat. $^nPr_4N^+RuO_4^-$, CH_2Cl_2, MS 3 Å, 0 °C

79% — 3. petroleum ether, **7**, 10 kbar, RT

15

Cy

B

Cy

7

Tips

- Some of the synthetic methods utilized have already been mentioned in this chapter.
- The first step is a *Sharpless* epoxidation of the allylic alcohol to a C-6, C-7 epoxide. Does application of L-(+)-dimethyltartrate lead to attack from a) below or b) above with respect to the structure as drawn?
- The functionality required for crotylation is achieved in the second step with $^nPr_4N^+RuO_4^-$ (TPAP).

- Which of the possible diastereomers arises during crotylation with reagent **7**?

Solution

TPAP is a standard reagent for oxidizing alcohols to aldehydes. It can be introduced either stoichiometrically or, as here, in catalytic amounts in combination with a cooxidant (NMO; see Chapter 16).

The absolute configurations of the new stereogenic centers in **15**, introduced through the *Sharpless* epoxidation and crotylation, can be predicted with the aid of the rules cited above. It is worth notice, however, that crotylation in the present case was carried out at 10 kbar. This is a frequently used modification of standard reaction conditions (with other synthetic methods as well) designed to improve selectivity and/or yield.

PMPh

15: R =

OH

Problem

PMPh

OH

7

6

15

72% 1. 1.0 eq ^{i}PrMgCl; 5.0 eq $LiAlH_4$, THF, 65 °C

97% 2. PMBCl, NaH, DMF, RT

16

Tips

- Addition of a *Grignard* reagent leads only to deprotonation of the alcohol at C-5 and thereby to intramolecular activation of the epoxide.
- Lithium aluminum hydride opens the epoxide ring reductively.
- The final protecting-group operation takes place selectively at the secondary alcohol.

Solution

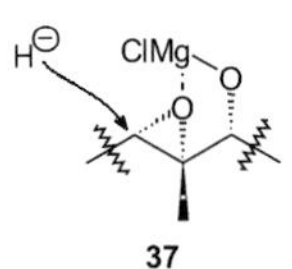

Especially interesting is the fact that the epoxide ring is activated by intramolecular chelation[25] (**37**). Prior to this discovery, 18 standard reagents had failed in opening the epoxide.[7] The reducing agent attacks activated epoxide **37** selectively from the sterically more accessible side and produces a tertiary alcohol. Selective protection of the secondary alcohol leads to compound **16**.

PMPh
O O O
PMBO
7 6
O HO
16

Problem

16 → (1. 2. 3. 4.) **17**

PMPh PMPh
O O O O O
7
9
O HO
17

Tips

- An analogous synthetic sequence was used to construct the C-7 through C-9 fragment.
- The sequence is: cleavage of the double bond, then crotylation, then protection.
- In this case there is need for one step more than with the C-7 through C-9 fragment.
- Cleavage of the double bond occurs in two steps.

Solution

The only difference in the synthetic sequence relative to the preparation of **12** consists of a variation in cleavage of the double bond. Employed here was a dihydroxylation followed by periodic acid treatment[26] (see Chapter 5) rather than the ozonolysis used with the C-7 through C-9 fragment.

O Cy
B
O Cy
7

1. NMO, cat. OsO_4, acetone, RT.
2. $NaIO_4$, THF, H_2O, 0 °C.
3. Petroleum ether, **7**, 10 kbar, RT, 70%.
4. DDQ, MS 3 Å, –20 °C, CH_2Cl_2, 82%.

Problem

Tips

- The acid function in **18** is derived from the double bond.
- To this end the double bond is first cleaved to an aldehyde.
- This takes place over two steps.
- After periodic acid cleavage of the double bond, the aldehyde is oxidized to an acid. What reagents are best suited to the task: a) CrO_3/H_2SO_4/acetone (*Jones* reagent); b) $NaClO_2/NaH_2PO_4$/2-methyl-2-butene; c) TPAP; d) PCC; or e) PDC?

Solution

Jones reagent is used to oxidize the aldehyde here to an acid. In general only the first two reagents listed in the "Tips" oxidize aldehydes to acids. TPAP can be used for oxidation of alcohols to aldehydes, ketones, or acids. The others are used for oxidizing alcohols to aldehydes or ketones.[27]

1. NMO, OsO_4 (cat.), acetone, 0 °C.
2. $NaIO_4$, THF, H_2O, RT.
3. CrO_3, H_2SO_4, acetone, –20 °C.

Problem

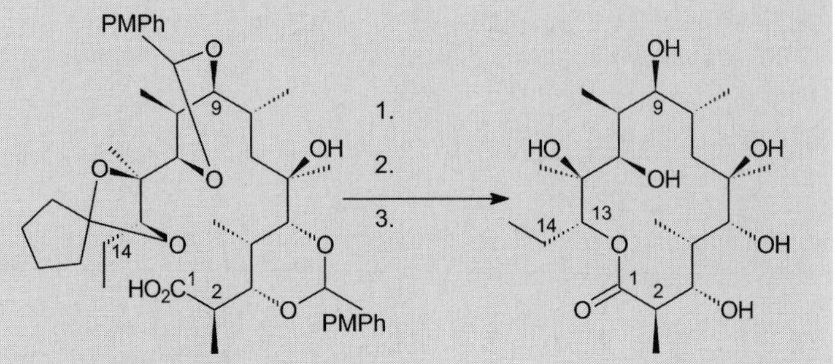

Tips

- In going from **18** to **19** the protecting groups are removed, but a macrolactonization is also achieved. In what order are these steps carried out?
- First the cyclopentylidene acetal is cleaved. What reaction conditions are used for cleaving acetals: a) acidic conditions, or b) basic conditions?
- This is followed by macrolactonization.
- For this purpose the carboxyl group is activated.
- Finally, the PMPh acetals are cleaved.

Solution

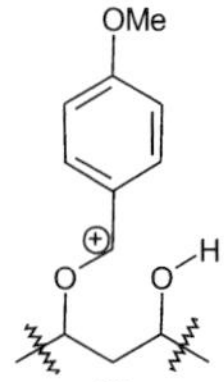

Although successfully carried out with model compounds, the cyclopentylidene acetal here could not be hydrolyzed selectively in the presence of the other protecting groups. Selective deprotection occurred only upon addition of trinitrotoluene, which as a more electron-deficient aromatic system forms a charge-transfer complex with the electron-rich aromatic portion of the PMPh acetal. The charge-transfer complex decreases the electron density of the acetal and thus increases its stability, presumably due to reduced stabilization of cation **38**. In this way it proved possible to release the alcohol functions at C-12 and C-13 selectively with acid.

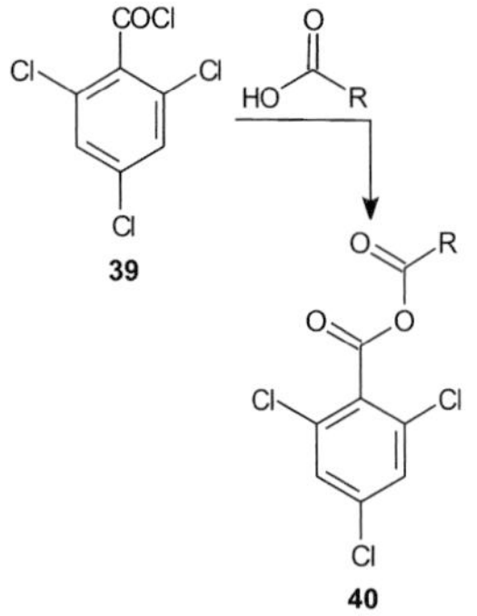

Macrolactonization of substrates such as **18** is a subject that has been investigated intensively. It was early established that the reaction proceeds well only if attention is paid to certain details.[1,5,28] First, the stereochemistry at C-9 is crucial, since only lactones with a 9*S* configuration are formed. For the same reason, the alcohol groups at C-3/5 and C-9/11 must be protected by cyclic systems. Furthermore, special activation of the carboxyl group as a thioester[29] or anhydride[30] is necessary. In this case the activation was introduced by the *Yamaguchi* method[31] through conversion into the anhydride **40** of trichlorobenzoic acid.

1. TNT (10 eq.), 2N HCl, CH_2Cl_2, MeOH, 35 °C.
2. Trichlorobenzoic anhydride (**39**), DMAP, toluene, RT.
3. 2N HCl, MeOH, 45 °C, 77%.

Solution

These steps constitute the end of *Hoffmann*'s synthesis, which thus terminates with (9*S*)-dihydroerythronolid A (**19**). Continuation of the synthesis to erythronolid A (**1**) had already been accomplished by *Kinoshita*.[31] Glycosidation of erythronolid A to erythromycin A (**2**) was carried out as early as 1981 by *Woodward*.[6] Thus, the synthesis presented by *Hoffmann* formally represents a total synthesis of erythromycin A (**2**).

Problem

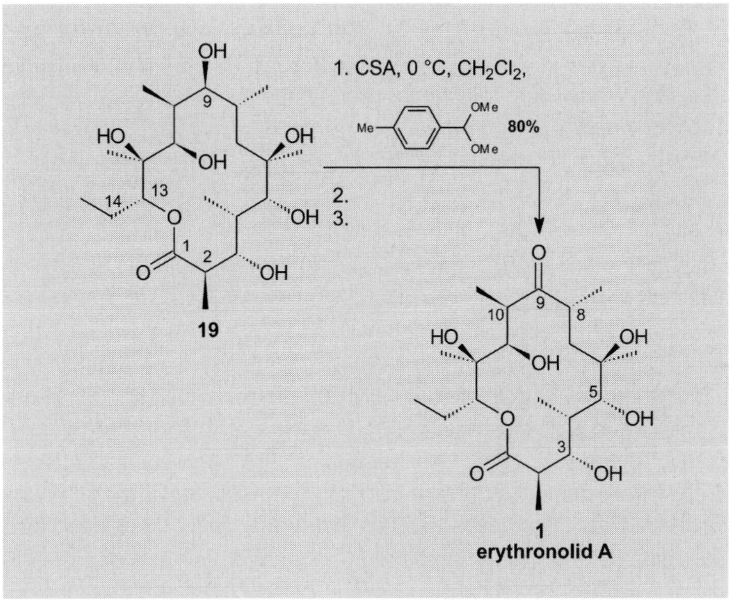

1. CSA, 0 °C, CH_2Cl_2, Me–C₆H₄–CH(OMe)₂, 80%
2.
3.

19 → **1** **erythronolid A**

Tips

- Acetalization under the conditions described is thermodynamically controlled.
- Aldehydes preferentially form acetals containing 6-membered rings, whereas ketones prefer 5-membered rings. Which of the two possible 6-membered-ring systems is thermodynamically favored?
- A decisive factor is the relative stereochemistry of the hydroxy groups.
- The OH group at C-9 can be selectively oxidized after protection of the alcohol functions at C-3 and C-5, although no simple explanation is apparent for this selectivity.

- The acetal is subsequently cleaved.
- Acid is avoided in this process to eliminate the risk of epimerization at C-8 and C-10.

Solution

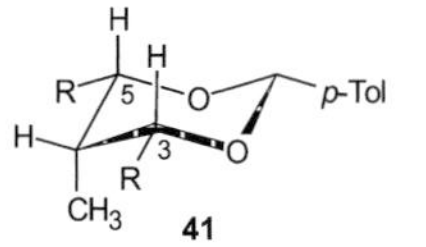

If molecule **19** is written out in a zigzag fashion, the hydroxy groups at C-9 and C-11 are seen to be *trans* to each other, whereas those at C-3 and C-5 are *cis*. In the chair conformation of the cyclohexane the two large substituents R of the C-3, C-5 acetal thus occupy equatorial positions (**41**), but this is not possible with the corresponding C-9, C-11 acetal **42**. For this reason the latter is unlikely to adopt a chair-like conformation. Even if these considerations represent only a first approximation, it is still possible to draw the correct conclusion, namely that the C-3,C-5 acetal **41** is thermodynamically favored, therefore leading to compound **43**.

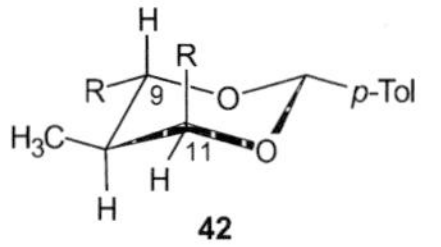

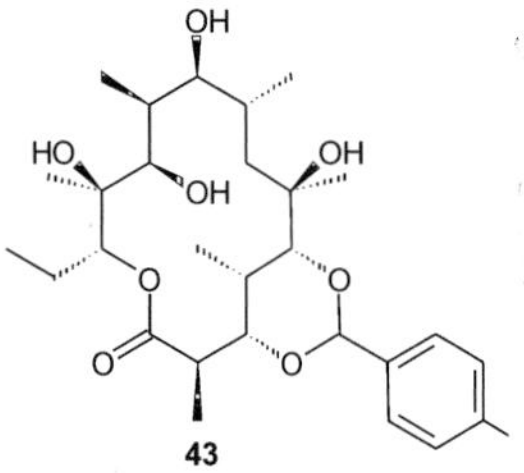

The hydroxy group at C-9 is later oxidized, after which the benzylidene acetal is cleaved by hydrogenolysis.[32] This eliminates any risk of epimerization that might accompany acidic cleavage.

1. PCC, MS 3 Å, 0 °C, 80%.
2. Pd/C, 1 bar H_2, MeOH, 82%.

6.4 Summary

The synthesis of erythronolid A presented here is may be linear, but it still represents the shortest known total synthesis of a compound in this category. Construction of the stereogenic centers is achieved exclusively through reagent control using the crotylation developed by *Hoffmann* together with *Sharpless* epoxidation. It thus proved possible to synthesize (9*S*)-dihydroerythronolid A (**19**) in an iterative procedure consisting of 23 steps with an overall yield of 10%.

6.5 References

1 J. Mulzer, *Angew. Chem. Int. Ed. Engl.* **1991**, *30*, 1452; *Angew. Chem. Int. Ed. Engl.* **1991**, *30*, 1452.

2 I. Paterson, M. M. Mansuri, *Tetrahedron* **1985**, *41*, 3569.

3 E. J. Corey et al., *J. Am. Chem. Soc.* **1978**, *100*, 4620.

4 R. B. Woodward et al., *J. Am. Chem. Soc.* **1981**, *103*, 3210.

5 R. B. Woodward et al., *J. Am. Chem. Soc.* **1981**, *103*, 3213.

6 R.B. Woodward et al., *J. Am. Chem. Soc.* **1981**, *103*, 3215.
7 R. Stürmer, K. Ritter, R.W. Hoffmann, *Angew. Chem. Int. Ed. Engl.* **1993**, *32*, 101; *Angew. Chem. Int. Ed. Engl.* **1993**, *32*, 101.
8 R. Stürmer, *Liebigs Ann. Chem.* **1991**, 311.
9 R.W. Hoffmann, R. Stürmer, *Chem. Ber.* **1994**, *127*, 2511; R. Stürmer, R.W. Hoffmann, *Chem. Ber.* **1994**, *127*, 2519.
10 Y. Gao, R.M. Hanson, J.M. Klunder, S.Y. Ko, H. Masamune, K.B. Sharpless, *J. Am. Chem Soc.* **1987**, *109*, 5765.
11 R.A. Johnson, K.B. Sharpless, *Catalytic Asymmetric Synthesis*, I. Ojima (Ed.), VCH, Weinheim, 1993, p. 103; see also Chapter 12.
12 A.J. Mancuso, D. Swern, *Synthesis* **1981**, 165.
13 R.W. Hoffmann, K. Ditrich, G. Köster, R. Stürmer, *Chem. Ber.* **1989**, *122*, 1783.
14 Further selective allylating and crotylating procedures: Y. Yamamoto, N. Asao, *Chem. Rev.* **1993**, *93*, 2207; R. Stürmer, R.W. Hoffmann, *Synlett* **1990**, 759; L.F. Tietze, A. Dölle, K. Schiemann, *Angew. Chem. Int. Ed. Engl.* **1992**, *31*, 1372; *Angew. Chem. Int. Ed. Engl.* **1992**, *31*, 1372; L.F. Tietze, K. Schiemann, C. Wegner, *J. Am. Chem. Soc.* **1995**, *117*, 5851; L.F. Tietze, K. Schiemann, C. Wegner, C. Wulff, *Chem. Eur. J.* **1996**, *2*, 1164; H.C. Brown, R.S. Randad, K.S. Bhat, M. Zaidlewicz, U.S. Racherla, *J. Am. Chem. Soc.* **1990**, *112*, 2389; A.L. Costa, M.G. Piazza, E. Tagliavini, C. Trombini, A. Umani-Ronchi, *J. Am. Chem. Soc.* **1993**, *115*, 7001; G.E. Keck, K.H. Tarbet, L.S. Geraci, *J. Am. Chem. Soc.* **1993**, *115*, 8467; K. Ishihara, M. Mouri, Q. Gao, T. Maruyama, K. Furuta, H. Yamamoto, *J. Am. Chem. Soc.* **1993**, *115*, 11490; S. Weigand, R. Brückner, *Chem. Eur. J.* **1996**, *2*, 1077; D.R. Gauthier Jr., E.M. Carreira, *Angew. Chem. Int. Ed. Engl.* **1996**, *35*, 2363; *Angew. Chem. Int. Ed. Engl.* **1996**, *35*, 2363.
15 T.W. Greene, P.G. M. Wuts, *Protective Groups in Organic Synthesis*, 2nd Edition, John Wiley & Sons Inc., New York 1991, p. 53.
16 P.J. Kocienski, *Protecting Groups*, Georg Thieme Verlag, Stuttgart 1994, p. 52.
17 W.R. Roush, *Comprehensive Organic Synthesis*, *Vol. 2* (Ed.: B.M. Trost, I. Fleming), Pergamon Press, Oxford, 1991, p. 1.
18 Y. Oikawa, T. Nishi, O. Yonemitsu, *Tetrahedron Lett.* **1983**, *24*, 4037.
19 E. Vedejs, M.J. Peterson, *Topics in Stereochemistry* **1994**, *21*, 1; B.E. Maryanoff, A.B. Reitz, *Chem. Rev.* **1989**, *89*, 863.
20 R. Brückner, *Reaktionsmechanismen*, Spektrum Akademischer Verlag, Heidelberg 1996, p. 318.
21 J. Boutagy, R. Thomas, *Chem. Rev.* **1974**, *74*, 87.
22 D.J. Ager, *Synthesis* **1984**, 384.
23 A.E.G. Miller, J.W. Biss, C.H. Schwartzman, *J. Org. Chem.* **1959**, *24*, 627.
24 E. Winterfeldt, *Synthesis* **1975**, 617.
25 J.A. Marshall, R.C. Andrews, *J. Org. Chem.* **1985**, *50*, 1602.
26 R. Pappo, D.S. Allen Jr., R.U. Lemieux, W.S. Johnson, *J. Org. Chem.* **1956**, *21*, 478.

27 R. Brückner, *Reaktionsmechanismen*, Spektrum Akademischer Verlag, Heidelberg 1996, p. 494; B.O. Lindgren, T. Nilsson, *Acta Chem. Scand.* **1973**, *27*, 888.

28 M. Hikota. H. Tone, K. Horita, O. Yonemitsu, *Tetrahedron* **1990**, *46*, 4613; G. Stork, S.D. Rychnovsky, *J. Am. Chem. Soc.* **1987**, *109*, 1565.

29 E.J. Corey, K.C. Nicolaou, *J. Am. Chem. Soc.* **1974**, *96*, 5614; E.J. Corey, D. Brunelle, *Tetrahedron Lett.* **1976**, *38*, 3409.

30 J. Inanaga, K. Hirata, H. Saeki, T. Katsuki, M. Yamaguchi, *Bull. Chem. Soc. Jpn.* **1979**, *52*, 1989.

31 M. Kinoshita, M. Arai, N. Ohsawa, M. Nakata, *Tetrahedron Lett.* **1986**, *27*, 1815.

32 T.W. Greene, P.G.M. Wuts, *Protective Groups in Organic Synthesis*, 2nd Edition, John Wiley & Sons Inc., New York 1991, p. 128; P.J. Kocienski, *Protecting Groups*, Georg Thieme Verlag, Stuttgart 1994, p. 96.

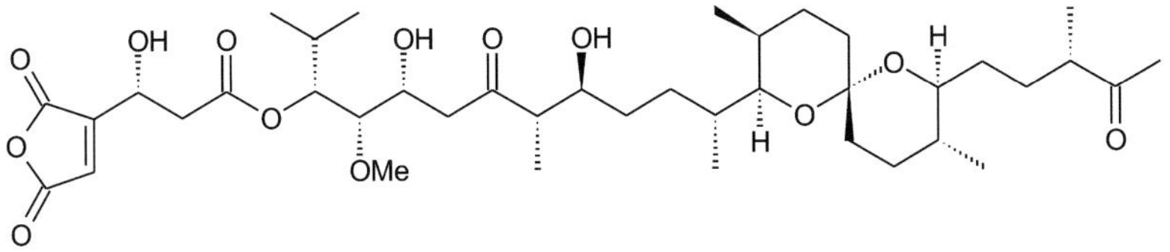

7

Tautomycin: Armstrong (1996)

7.1 Introduction

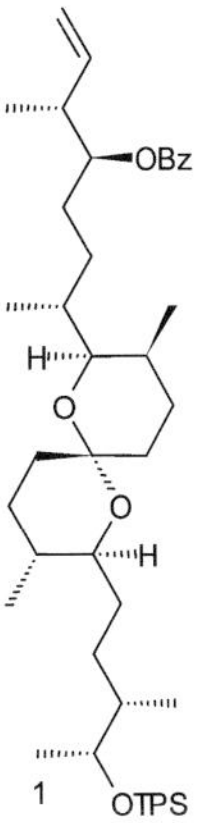

Tautomycin, a secondary metabolite from *Streptomyces spiroverticillatus*, was first isolated by *Isono*, who also established its structure.[1] The unusual structural feature of this compound is an unsaturated anhydride group, which in aqueous medium is in equilibrium with the free acid. This probably is essential to the compound's biological activity, as is suggested by its structural relationship to such substances as the calyculins[2] or okadaic acid.[3]

All these compounds are selective inhibitors of the serine/threonine phosphatases PP1 and PP2A.[4] Phosphatases catalyze the hydrolysis of phosphates bound to serine or threonine OH groups in enzymes, and hydrolysis of this type causes the enzymes to be activated or deactivated. Among other functions these enzymes play a role in the regulation and control of glycogen metabolism.

Isono's group succeeded in establishing the absolute configurations of all 13 stereocenters[5] of tautomycin in 1993 through chemical degradation combined with NMR analysis and force-field calculations. NMR analysis not only established the various coupling constants but also permitted determination of absolute configurations of the individual centers with the aid of derivatives incorporating mandelic and *Mosher*'s acids.

Several partial syntheses of tautomycin have been reported, including the synthesis of a C-1 through C-26 fragment by *Shibasaki*,[6] followed by a first total synthesis in 1994 by *Oikawa*.[7] In 1997 *Isobe*[8] presented a convergent total synthesis starting from two building blocks.

In this chapter we describe a partial synthesis of the C-1 through C-21 fragment of **1** devised by *Armstrong*.[9] One striking feature is the iterative use of a method by which functionalized alkyl chains are constructed stereoselectively.

21 OBz H 1 O O H 1 OTPS

7.2 Overview

HO
HO
H
CH_3
2
1.
2.
H
B
O
O
Cl
CH_3
3
1. MgBr 13
THF, - 78 °C, 80%
2. CH_2Cl_2, ^{n}BuLi, THF
-100 °C, $ZnCl_2$, 92%
3. THF, DMSO
-78 °C - RT, 76%
MeO
OLi
14
4
1.
2.
3.
H
Cl
B
O
O
CH_3
OTBS
OPMB
5
NaHB(OMe)$_3$
THF, RT
6
1.
2.
3.
OTBS
OPMB
7
1.
2.
3.
OTBS
OBz
OPMB
8
1.
2.
O
OBz
OPMB
9

OBz OPMB
9
+
I OPMB OTPS
10

OBz OPMB OH OPMB OTPS
11

1.
2.

OBz OPMB O OPMB OTPS
12

OBz H O O H OTPS
1
tautomycin-fragment

7.3 Synthesis

Problem

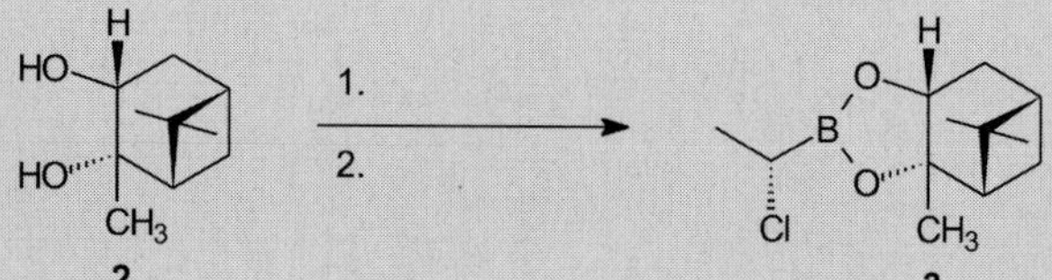

HO, HO, H, CH_3, **2** → (1., 2.) **3** (B, O, O, H, Cl, CH_3)

Tips

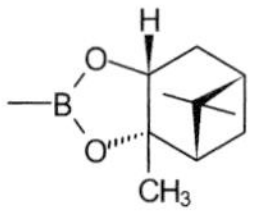

15

- To what category of compounds does **3** belong? Is it a) an ether; b) an ester; c) an acid; or d) an acetal?
- Compound **3** is an ester of a boronic acid!
- The first step is a condensation reaction with methanoboronic acid to give **15**. How many carbon atoms must still be introduced?
- In this C_1 extension process, a nucleophile attacks the electrophilic boron atom. What results is an *at* complex.
- Which combination of reagents leads to the *at* complex that then decomposes to the product: a) $CHCl_3$, KO^tBu; b) CH_3Cl, Mg; c) CH_2Cl_2, ^nBuLi; d) $CHCl_3$, $CrCl_2$?

Solution

$^nBuLi + CH_2Cl_2 \rightarrow$

$^nBuH + LiCHCl_2$

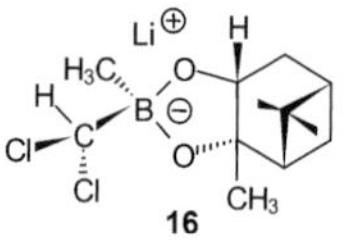

16

Dichloromethane is first deprotonated with ^nBuLi in tetrahydrofuran at –100 °C. The reaction of dichloromethane with ^nBuLi represents a competition between deprotonation and halogen-metal exchange. When reaction is carried out at very low temperature only deprotonation occurs. The *at* complex **16** forms after addition of the methanoboronic acid pinanediol ester **15**.[10] Upon introduction of $ZnCl_2$ this complex rearranges to compound **3**.[11]

The mode of action of $ZnCl_2$ has not been clarified. Nevertheless, it is thought to interact with complex **16**, because its presence increases the rate of reaction. Moreover, it seems to be involved in cleavage of one of the two diastereotopic chlorine atoms, because the diastereoselectivity of the reaction increases simultaneously.

1. Methanoboronic acid, Na_2SO_4, Et_2O, 94%.
2. ^nBuLi, CH_2Cl_2, THF, –100 °C; $ZnCl_2$, 61%.

The mechanism of the rearrangement is as follows:[12]

16 *at* complex → ($ZnCl_2$) → 3

Discussion

It is important to use THF as solvent, because this avoids α elimination[13] to a monochlorocarbene. The phenomenon is explained by the observation that a lithium cation, acting as a Lewis acid, facilitates halogen cleavage and the formation of LiCl. THF complexes with the cation very effectively and thereby reduces its electrophilicity.

$^nBuLi + CH_2Cl_2 \rightarrow$

$^nBuH + LiCl + :ChCl$

Problem

3 → 4

1. MgBr 13, THF, -78°C, **80%**
2. CH_2Cl_2, ^nBuLi, THF, -100°C, $ZnCl_2$, **92%**
3. THF, DMSO, -78°C - RT, **76%**, MeO–C₆H₄–CH₂OLi **14**

Tips

- In the first step, the *Grignard* reagent reacts with the substrate as a nucleophile. This is not an S_N2 reaction.
- The reaction itself has already been introduced.
- The result is another *at* complex.
- The *at* complex rearranges with substitution of a chlorine atom.
- In a second step involving the reagent combination CH_2Cl_2/$^nBuLi/ZnCl_2$, the alkyl chain is once again extended by one chloromethylene unit.
- The stereochemistry is the same as in the case of product **3**.
- A third step causes the lithium salt of *p*-methoxybenzyl alcohol **14** to be subjected to the same reaction sequence as the two previous reagents.
- The *at* complex again rearranges with substitution of a chlorine atom.

Solution

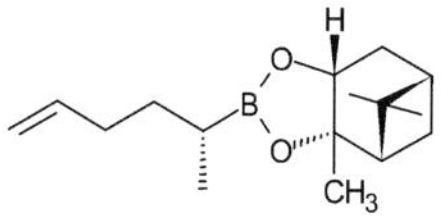

17

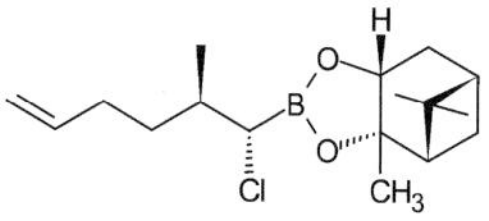

18

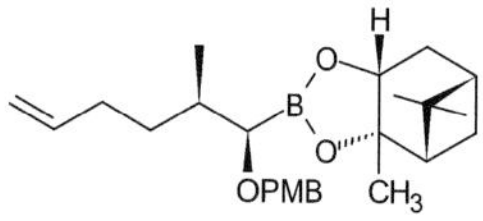

4

The first step is nucleophilic addition of *Grignard* reagent **13** to the electrophilic boron atom of boronic acid ester **3** with formation of *at* complex **19**. This rearranges to **17** by substitution of a chlorine atom. No supplemental Lewis acid is required for this rearrangement. Another halogenated building block is then introduced in the previously described way by the addition of dichloromethyllithium (CH_2Cl_2 + ^nBuLi) with formation of **18**. This is subsequently functionalized with the lithium salt of *p*-methylbenzyl alcohol **14**. The formation of *at* complexes is not limited to carbon nucleophiles: **14** also adds in an analogous way and rearranges to **4**.

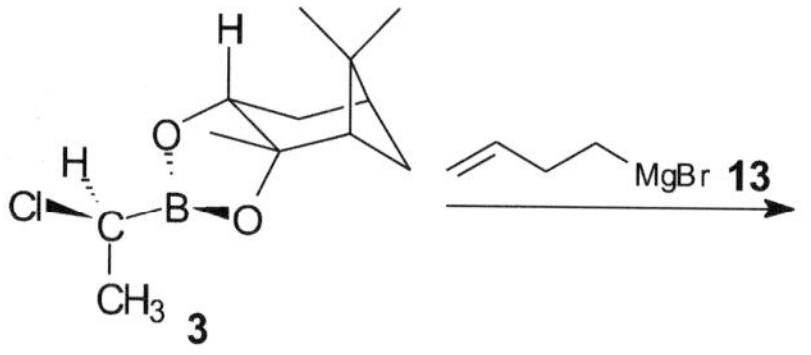

MgBr **13**

3

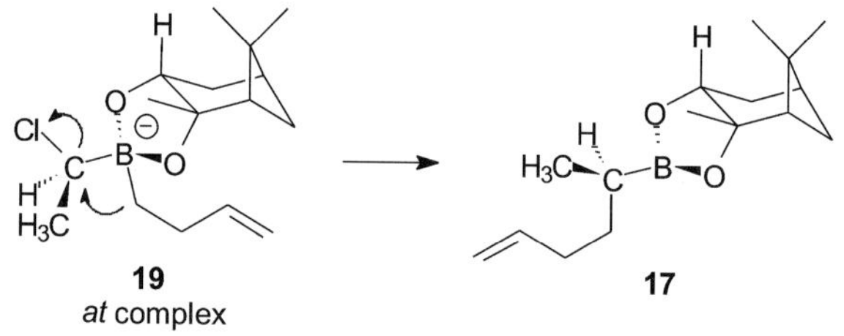

19
at complex

17

Problem

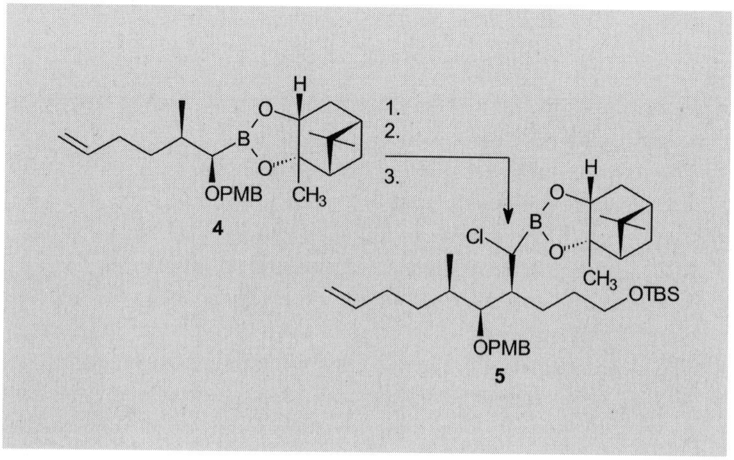

1.
2.
3.

4

5

Tips

- By how many carbon atoms has the distance from the boron atom to the PMB group changed?
- The distance has increased to three carbon atoms. Therefore, how many times has an insertion reaction with dichloromethyllithium occurred?
- The first and third steps are insertion reactions involving dichloromethyllithium.
- The second step entails addition of another carbon nucleophile.
- A functionalized C_3 residue is introduced in the form of a *Grignard* reagent in the course of the reaction sequence with formation of an *at* complex and rearrangement.

Solution

Application of the previously described reaction sequence allows an initial insertion reaction with dichloromethyllithium to give **20**. Then **20** is enhanced by one C_3 residue using *Grignard* reagent **21** to produce **22**. Another reaction with dichloromethyllithium results in a C_1 chain extension to product **5**. Since the insertion reaction with dichloromethyllithium was conducted twice, the distance between the boron atom and the PMB ether increases by two carbon atoms.

1. CH_2Cl_2, ^nBuLi, THF, –100 °C; $ZnCl_2$, 92%.
2. **21**, THF, –78 °C → RT, 71%.
3. CH_2Cl_2, ^nBuLi, THF, –100 °C; $ZnCl_2$, 49%.

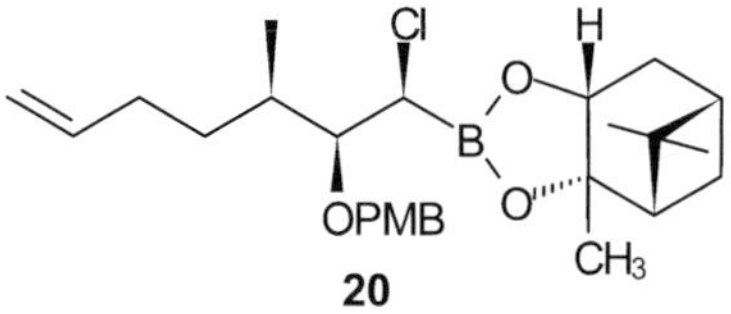

BrMg OTBS
20

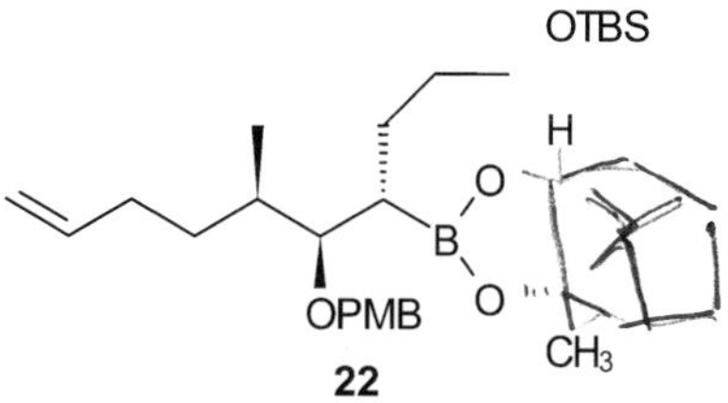

OTBS Cl H O B O OPMB CH3 20 H O B O OPMB CH3 22

Discussion

The yield in the last insertion reaction is only 49%, because byproduct **23** is formed in 34% yield. Notice that in **23** the double bond is at precisely the spot where previously there were bonds to the boronic acid ester and the PMB ether. It is thus reasonable to assume that the structural elements consisting of the boronic acid ester and the PMB ether have a tendency to undergo elimination under the influence of base.

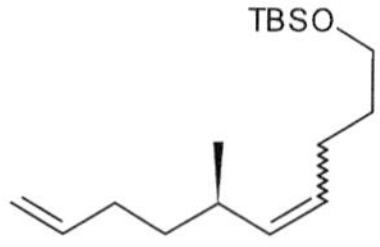

TBSO
23

Problem

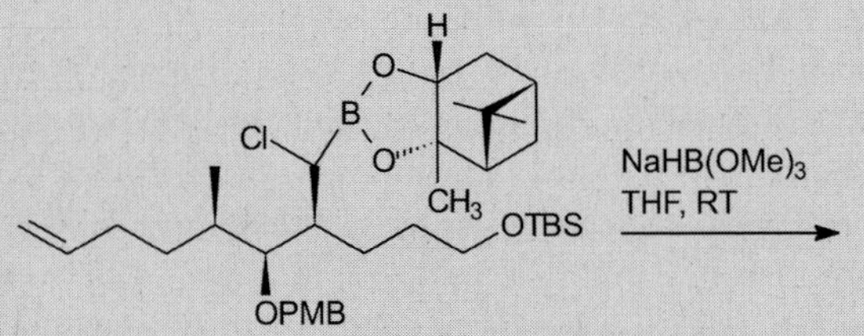

Tip

- A hydride ion from the boron reagent substitutes selectively for one functional group in **5**.

Solution

The chlorine atom is replaced by hydride. The actual reagent forms in the course of adding $NaBH_4$ to methanol, and it is frequently employed as a mild reducing agent. The reduced reactivity of this reagent relative to $NaBH_4$ is a result of the +M effect of the alkoxy group and its increased steric demand. Greater selectivity is thus observed. Compound **6** is carried into the next reaction as a crude product.

Problem

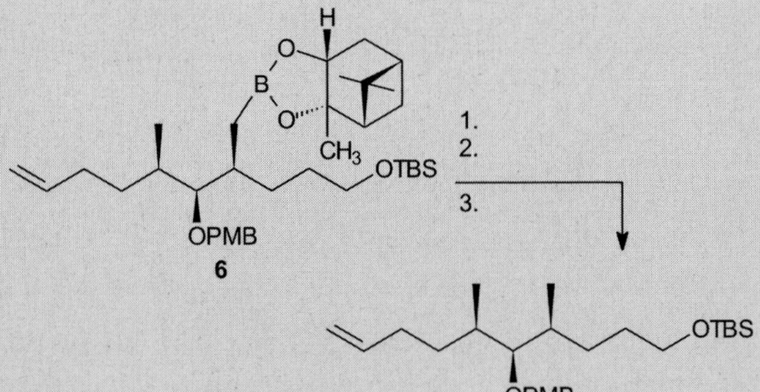

Tips

- The first step utilizes the same reagents as for an oxidative workup after a hydroboration (see Chapter 5).
- An alcohol is produced.
- This alcohol is transformed in the second reaction into a mesylate.
- In the third step, is the mesylate converted into a methyl group by oxidation or reduction?

Solution

Boronate **6** is oxidized with alkaline hydrogen peroxide to alcohol **24**. The corresponding mesylate is reduced with lithium aluminum hydride to a methyl group.

1. NaOH, H_2O_2, THF, RT, 87% starting from **5**.
2. MsCl, NEt_3, CH_2Cl_2, 0 °C, 94%.
3. $LiAlH_4$, THF, 60 °C, 88%.

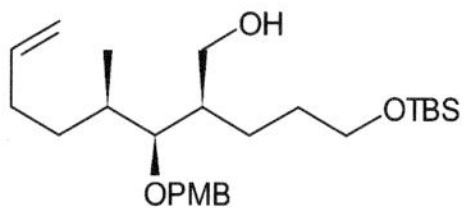

OH
OTBS
OPMB
24

Discussion

A plausible alternative to oxidative cleavage of the boronate ester is protolysis with acid.[14] This is perfectly appropriate for trialkylboranes, but monoalkylboranes and boronate esters are cleaved only under more drastic conditions. Experiment showed that these conditions are too severe for the substrate, requiring the detour by way of an oxidation/deoxygenation sequence.

Problem

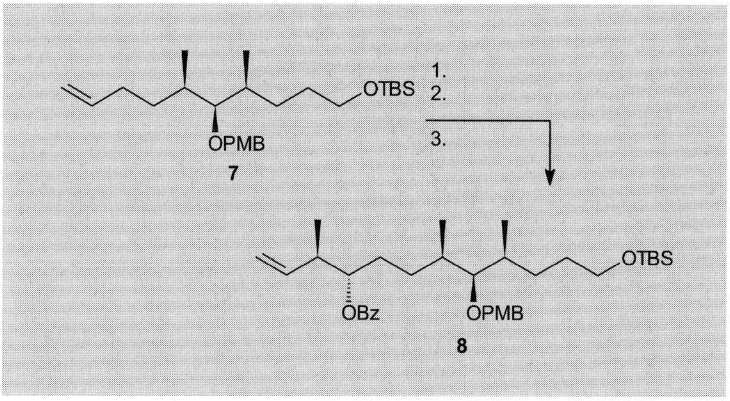

OTBS
OPMB
7
1.
2.
3.
OTBS
OBz
OPMB
8

Tips

- What transpires here is a C_4 chain extension with simultaneous creation of two stereogenic centers.
- A homoallylic alcohol is synthesized (see Chapter 6).
- A crotylboron reagent is employed.
- In the first step, compound **7** is shortened by one carbon atom.
- A reagent containing three oxygen atoms is used to convert an olefin into an aldehyde.
- Preparation of the benzoate occurs in the last step.

Solution

The double bond in **7** is cleaved to an aldehyde by ozonolysis (see Chapter 5) and subsequent reductive workup with tributylphosphine. Crotylation with reagent **25**,[15] developed by *Brown*, produces only a single diastereomer, which is then transformed into benzoate **8** with benzoyl chloride in pyridine.

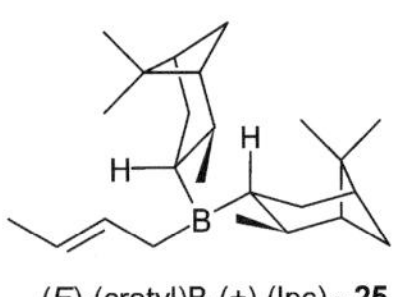

(E)-(crotyl)B-(+)-(Ipc)$_2$ **25**

1. O_3, CH_2Cl_2, –78 °C; PBu_3, 93%.
2. **25**, THF, –78 °C, 71%.
3. BzCl, DMAP, pyridine, RT, 85%.

Problem

OTBS 1.
2.
OBz OPMB
8
OBz OPMB
9

Tips

- The first step is a protecting-group operation.
- What type of reaction is employed to prepare the aldehyde: a) reduction; b) S_N2 reaction; or c) oxidation?

Solution

The silyl protecting group is cleaved with TBAF in THF, and the resulting free alcohol is then converted into aldehyde **9** by a *Swern* oxidation.

1. TBAF, THF, 0 °C, 100%.
2. DMSO, $(COCl)_2$, CH_2Cl_2, NEt_3, –78 °C, 69%.

The next reaction accomplishes coupling of the building blocks **9** and **10**. Fragment **10** is prepared through iterative use of (R)-pinanediol, the enantiomer of **2**, by an addition-functionalization sequence similar to that already presented. The vinylic iodide is prepared with a *Takai* reaction (see Chapter 13).

Problem

Tips

- The reaction of vinylic iodide **10** with aldehyde **9** occurs by way of an organometallic intermediate.
- This is not a *Grignard* reaction.
- Chromium(II) chloride is employed, and the addition of nickel(II) chloride has a catalytic effect.
- Chromium(II) reduces the nickel(II) to nickel(0) and itself becomes chromium(III).
- Nickel(0) adds oxidatively to vinylic iodide **10**.
- The resulting intermediate undergoes transmetallation with one of the two chromium species, and it then reacts with the aldehyde.

Solution

The reaction utilized here for coupling fragments **9** and **10** is called the *Nozaki–Hiyama* reaction,[16] and consists of adding an alkenyl chromium compound to an aldehyde with the formation of an allylic alcohol.

In the course of his synthesis of palytoxin, *Kishi*[17] noted that only certain batches of $CrCl_2$ were effective. Closer investigation showed that an impurity of $NiCl_2$ is essential if the reaction is to succeed.

The following mechanism has been postulated: Cr(II) reduces Ni(II) to Ni(0), which undergoes oxidative addition to vinylic iodide **26**. In the process, the Ni(0) species is transformed into a Ni(II) compound, which is transmetallated by the resulting Cr(III) species to give **28**. This compound then adds to the aldehyde, leading to allylic alcohol **29**.

This is illustrated below.

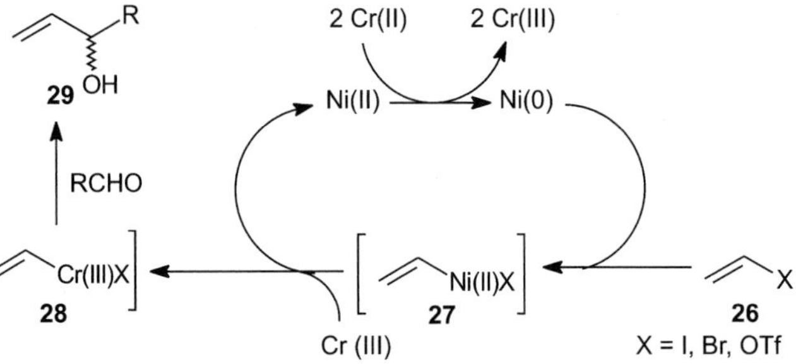

$CrCl_2$, $NiCl_2$ (cat.), THF, DMF, RT, 65%.

Discussion

This reaction is chemoselective, because only the aldehyde function is attacked even in the presence of other electrophilic groups such as ketones or nitriles.[18]

Chromium adds oxidatively not only to vinylic iodides, but also to allylic, aryl, and alkynyl iodides, as well as to CHI_3 (see Chapter 13).[18] The corresponding triflates can be employed equally well. On the other hand, the reactivity of bromides and chlorides is usually too limited. Reaction is carried out in the polar aprotic solvent DMF because this is capable of dissolving both of the salts, and homogeneous conditions accelerate the reaction.

Problem

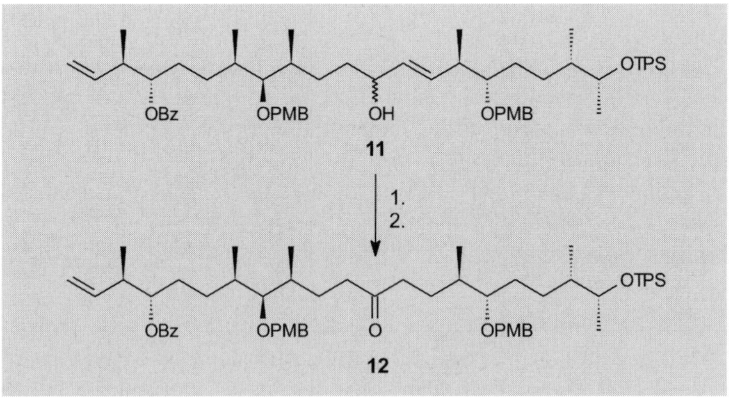

Tips

- One of the steps is a reduction, the other an oxidation.
- Which combination of reagents is appropriate: a) *Swern* oxidation, $NaBH_4$; b) $[Ir(COD)py(PCy_3)PF_6]$, H_2; $KMnO_4$; c) Pd/C/H_2; CAN; d) *Dess-Martin* periodinane; $[PPh_3CuH]_6$; or e) MnO_2, $Na_2S_2O_4$?
- The first step is oxidation to an α,β-unsaturated ketone.
- What characteristics must a reagent possess if it is to give exclusively 1,4 reduction?

Solution

For oxidation to an α,β-unsaturated ketone in this case the reagent of choice was the *Dess-Martin* periodinane **30**.[19] The copper hydride complex $[PPh_3CuH]_6$ then leads to saturated ketone **12**.[20]

1. *Dess-Martin* periodinane, CH_2Cl_2, 96%.
2. $[PPh_3CuH]_6$, benzene, 99%.

Discussion

Oxidation with *Dess-Martin* reagent **30** is noteworthy for its simple preparative methodology as well as its efficiency. The process involves a hypervalent iodine reagent that can be prepared easily and in high yield in two steps from *o*-iodobenzoic acid. Two equivalents of acetic acid are formed per mole of alcohol, which means compounds that are highly sensitive to acid cannot be oxidized in this way.

The mechanism of the *Dess-Martin* oxidation is as follows:[21]

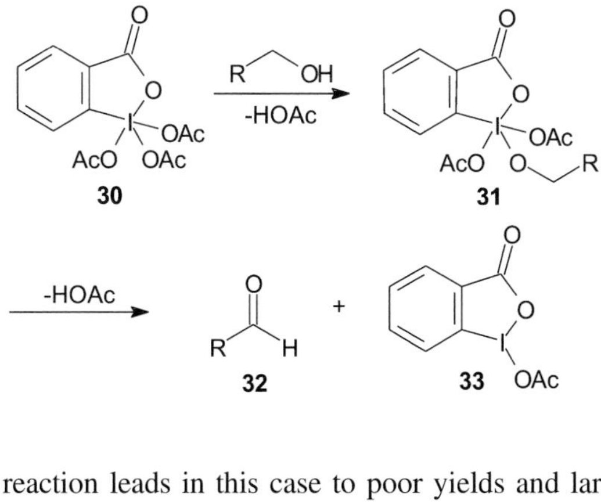

A *Swern* reaction leads in this case to poor yields and large quantities of decomposition products.

Among the other reagents listed, activated MnO_2 is specific for the oxidation of allylic and benzylic alcohols.[22] $KMnO_4$ is used for the oxidation of alkenes to diols.[23] Ceriv(IV) ammonium nitrate (CAN) accomplishes oxidative cleavage of the PMB protecting group.

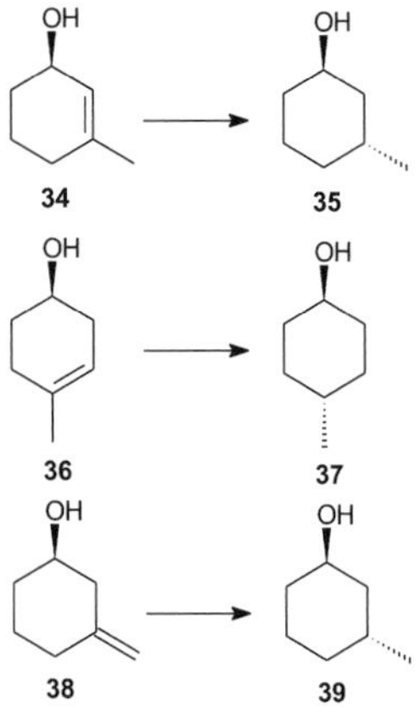

Reduction to a saturated ketone raises the potential problem of competition between 1,2 and 1,4 reduction. Based on the HSAB principle, soft nucleophiles (e.g., cuprates) lead to 1,4 products. On the other hand, hard nucleophiles (alkyllithium compounds, $LiAlH_4$) react preferentially to give 1,2 products. Two alternatives to the copper hydride complex $[PPh_3CuH]_6$ were investigated – reduction with either $NaBH_3$ or $Na_2S_2O_4$[24] – but both failed to produce the desired results.

Catalytic hydrogenation of the double bond of the allylic alcohol with either Pd/C or an iridium catalyst would lead to reduction also of the terminal olefinic linkage.

The iridium catalyst $[Ir(COD)py(PCy_3)PF_6]$ is utilized in substrate-controlled hydrogenations of allylic alcohols of the type **34**, **36**, and **38**.[25] The catalyst coordinates with the hydroxy group in the molecule, so hydrogenation occurs only from one side.

Problem

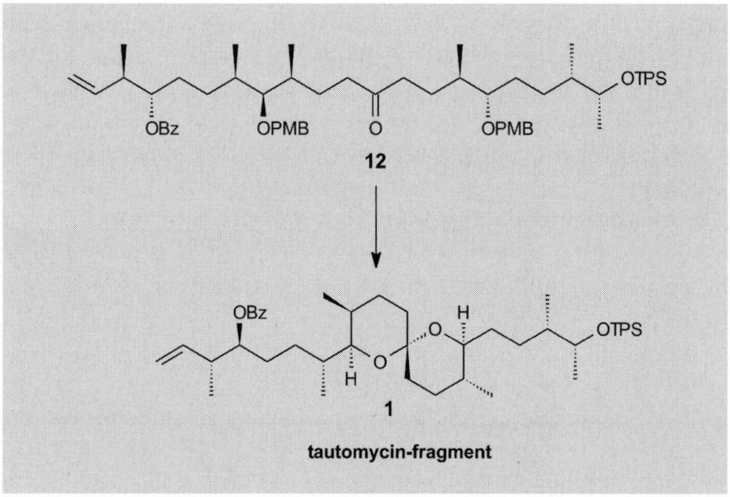

tautomycin-fragment

Tips

- Something is removed from the molecule to permit spiroacetalization to occur.

Solution

The PMB protecting group is cleaved with DDQ. Both free hydroxy groups then react with the ketone in a condensation reaction to give spiroacetal **1**.
DDQ, CH_2Cl_2, RT, 67%.

7.4 Summary

The synthesis described above is distinctive for its iterative use of a reaction sequence consisting of insertion and subsequent functionalization of a chloromethylene group. The advantage of this sequence is that the chain grows on the auxiliary reagent. Each new insertion reaction of dichloromethyllithium thereby occurs with high diastereoselectivity relative to the remaining chlorine atom. However, the stereochemistry of the insertion reaction itself is determined by which isomer of the auxiliary is employed. This addition/rearrangement reaction of a nucleophile permits a broad spectrum of substituents to be introduced.

The convergent nature of the process starting from **2** means that the number of steps has also been limited to 18. Joining of the two building blocks **9** and **10** occurs via the *Nozaki-Hiyama* reaction. Ketone **12** is obtained through an oxidation-reduction sequence starting with allylic alcohol **11**.

The chief reason for utilizing the reaction sequence described is the fact that formation of the α,β-unsaturated ketone in **11** makes it possible to differentiate between the two double bonds, because the copper hydride complex $[PPh_3CuH]_6$ reduces only the conjugated double bond.

It is worth considering a bit the protecting-group strategy that has here been employed. The double bond introduced at the beginning of the synthesis (**3** → **4**) is not activated in the form of an aldehyde equivalent until the transformation of **7** into **8**. This is an example of the "place-holder concept."[26] In this way alkenes or protected alcohols – which are simpler and can be protected in various ways – are introduced in the course of a synthesis as equivalents for sensitive carbonyl functions.

The two missing stereogenic centers are supplied under reagent control by crotylation.

Formation of a benzoate represents introduction of a protecting group of a new type (an ester). A distinguishing feature of building block **8** is the use of protecting groups with orthogonal stability[26] (see Chapter 13). That is to say, each of the three protecting groups in the molecule can be cleaved independently with a different combination of reagents (silyl: fluoride; PMB: oxidatively with DDQ or CAN; ester: alkaline).

Building block **10** also contains a PMB protecting group. The resulting simplification of the protection for the hydroxy groups required for spiroacetalization permits simultaneous cleavage with the formation of **1**.

7.5 References

1 X.-C. Cheng, T. Kihara, H. Kusakabe, J. Magea, Y. Kobayashi, R.-P. Fang, Z.-F. Ni, Y.-C. Shen, K. Ko, I. Yamaguchi, K. Isono, *J. Antibiot.* **1987**, *40*, 907.

2 S. Matsunaga, H. Fujiki, D. Sakata, N. Fusetani, *Tetrahedron* **1991**, *47*, 2999.

3 K. Tachibana, P.J. Scheurer, Y. Tsukitani, H. Kikuchi, D. Van Engen, J. Clardy, Y. Gopichand, F.J. Schmitz, *J. Am. Chem. Soc.* **1981**, *103*, 2469.

4 E.G. Krebs, *Angew. Chem. Int. Ed. Engl.* **1993**, *32,* 1122; E.H. Fischer, *Angew. Chem. Int. Ed. Engl.* **1993**, *32,* 1130.

5 M. Ubukata, X.-C. Cheng, M. Isobe, K. Isono, *J. Chem. Soc. Perkin Trans.* **1993**, 617.

6 S. Nakamura, M. Shibasaki, *Tetrahedron Lett.* **1994**, *35*, 4145.

7 H. Oikawa, M. Oikawa, T. Ueno, A. Ichihara, *Tetrahedron Lett.* **1994**, *35*, 4809.

8 K. Tsuboi, Y. Ichikawa, Y. Jiang, A. Naganawa, M. Isobe, *Tetrahedron* **1997**, *53*, 5123.

9 K.W. Maurer, R.W. Armstrong, *J. Org. Chem.* **1996**, *61*, 3106.

10 D.S. Matteson, R. Ray, R.R. Rocks, D.J. Tsai, P.K. Jesthi, D.S. Matteson, *Organometallics* **1983**, *2*, 1543.

11 D.S. Matteson, D. Majumdar, *Organometallics* **1983**, *2*, 1529.

12 S.E. Gibson nee Thomas, *The roles of boron and silicon*, Wiley-VCH, Weinheim 1995.

13 G. Köbrich, H.R. Merkle, H. Trapp, *Tetrahedron Lett.* **1965**, *15*, 969.

14 H.C. Brown, N.C. Herbert, *J. Organomet. Chem.* **1983**, *255*, 135.

15 H.C. Brown, P.K. Jadhav, S.K. Baht, *J. Am. Chem. Soc.* **1988**, *110*, 1535.

16 Kishi, *Pure Appl. Chem.* **1992**, *64*, 343; K. Takai, M. Tagashira, T. Kuroda, K. Oshima, K. Utimoto, H. Nozaki, *J. Am. Chem. Soc.* **1986**, *108*, 6048.

17 E.M. Suh, Y. Kishi, *J. Am. Chem. Soc.* **1994**, *116*, 11205.

18 P. Cintas, *Synthesis* **1992**, 248.

19 R.E. Ireland, L. Liu, *J. Org. Chem.* **1993**, *58*, 2899.

20 D.M. Brestensky, J.M. Stryker, *Tetrahedron Lett.* **1989**, *30*, 5677.

21 D.B. Dess, J.C. Martin, *J. Am. Chem. Soc.* **1991**, *113*, 7277.

22 A.J. Fatiadi, *Synthesis* **1976**, 133.

23 M.B. Smith, *Organic Synthesis*, McGraw-Hill New York, 1994,P. 281.

24 O. Louis-Andre, G. Gelbard, *Tetrahedron Lett.* **1985**, *26*, 831.

25 A.H. Hoveyda, D.A. Evans, G.C. Fu, *Chem. Rev.* **1993**, *93*, 1307.

26 M. Schelhaas, H. Waldmann, *Angew. Chem. Int. Ed. Engl.* **1996**, *35*, 2056.

8

(–)-α-Thujone: Oppolzer (1997)

8.1 Introduction

The thujones are found in true wormwood, *Artemisia absinthum*, a species of mugwort native to Europe and Asia. This plant was once a source for absinthe, a greenish, bitter-tasting alcoholic beverage that has since been banned because of the toxicity of the thujones. The compounds have a peppermint-like odor and act as nerve poisons that can be the cause of epileptic fits. Nevertheless, they still find some application in homeopathic medicine.

Apart from (–)-α-thujone (**1**), other stereoisomers such as (+)-β-thujone (**2**) and reduced derivatives like 3-neoisothujanol (**3**) can be isolated from appropriate extracts of wormwood, arbor vitae, and sage. An oil derived from leaves of the red cedar *Thulia plicata Don* contains 80–90% (–)-α-thujone (**1**).[1]

Compounds **1**–**3** are distinctive for their striking bicyclo[3.1.0]hexane skeleton, which was prepared in enantiomerically pure form for the first time in the synthesis described here, a linear process involving a palladium-catalyzed domino enyne cyclization[2] as the key reaction.

Earlier preparative work associated with the thujones was primarily directed toward partial synthesis and to derivatization of the natural products in the course of structure proof.[3]

(-)-α-thujone
1

(+)-β-thujone
2

3-neoisothujanol
3

8.2 Overview

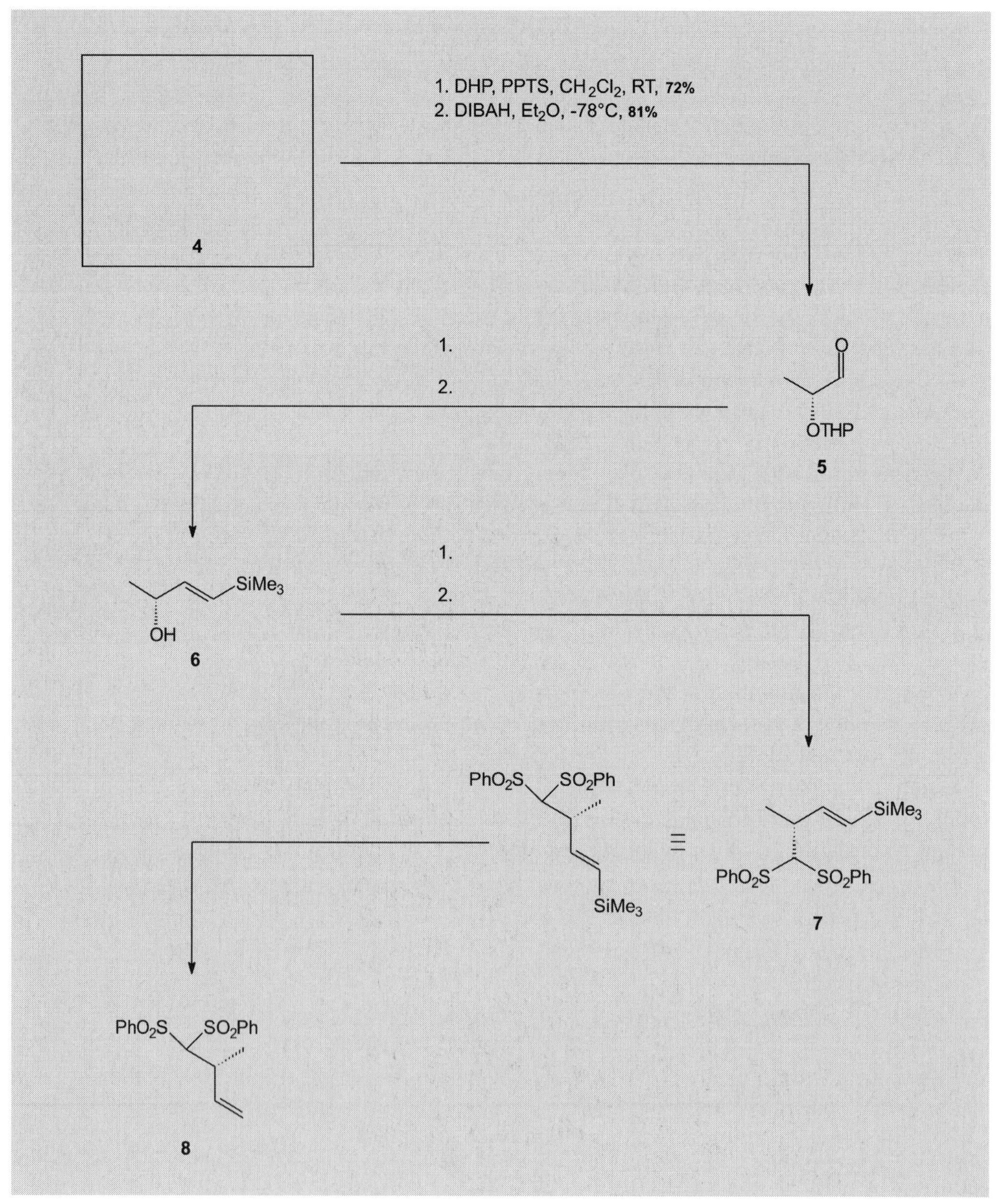
1. DHP, PPTS, CH2Cl2, RT, 72%
2. DIBAH, Et2O, -78°C, 81%
4
1.
2.
OTHP
5
SiMe3
OH
6
1.
2.
PhO2S
SO2Ph
SiMe3
≡
SiMe3
PhO2S
SO2Ph
7
PhO2S
SO2Ph
8

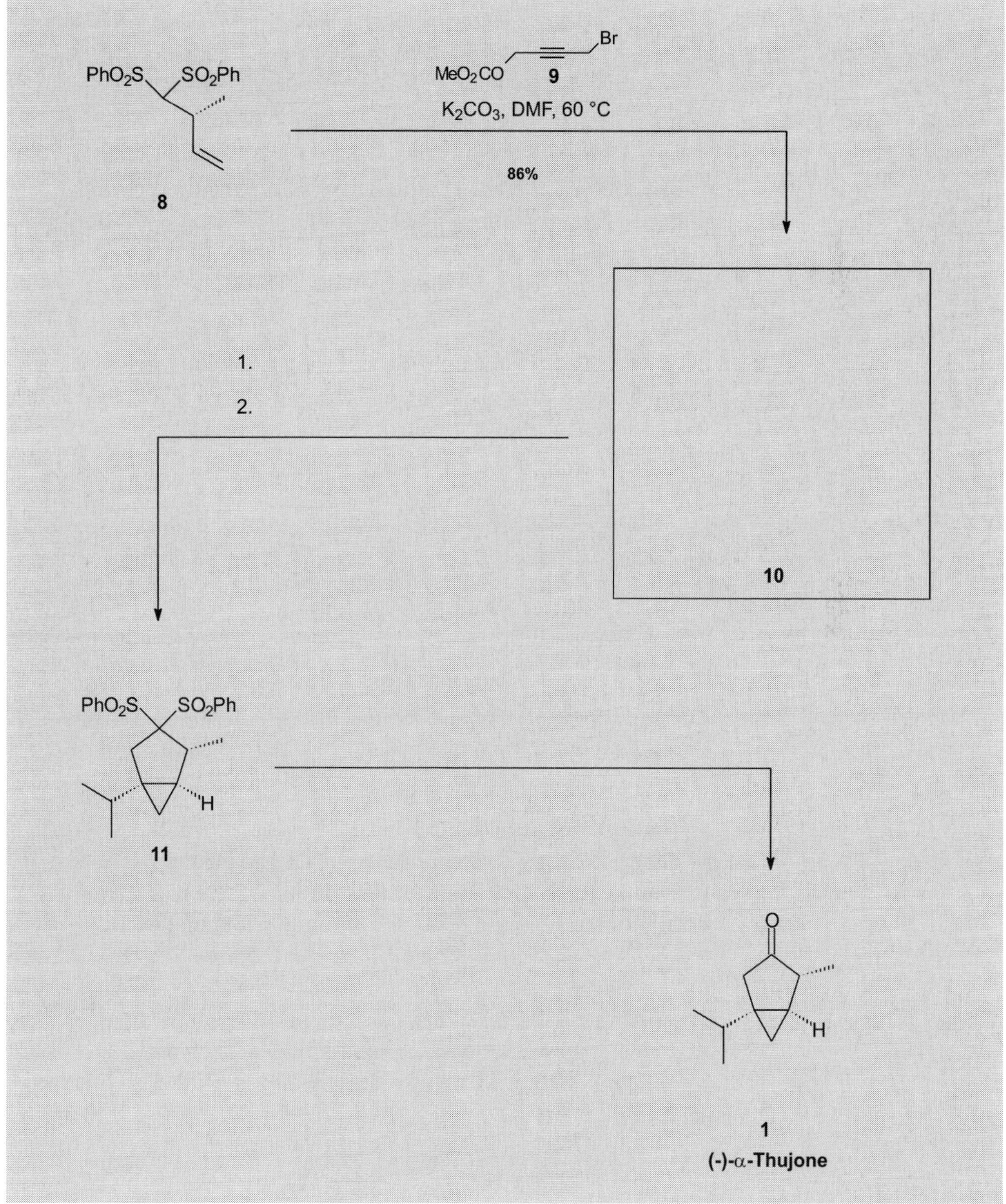
PhO2S SO2Ph
8
MeO2CO
Br
9
K2CO3, DMF, 60 °C
86%
10
1.
2.
PhO2S SO2Ph
H
11
O
H
1
(-)-α-Thujone

8.3 Synthesis

Problem

1. DHP, PPTS, CH_2Cl_2, RT, 72%
2. DIBAH, Et_2O, -78 °C, 81%

4 → 5 (OTHP)

Tips

- The starting material is a derivative of a natural product.
- In the first reaction, a hydroxy group is reacted with dihydropyran.
- The second step is a reduction with DIBAH.

Solution

OEt, OH, 4

Ethyl lactate (**4**) is transformed in the first step into the corresponding THP-protected derivative, and this is reduced with DIBAH in a second reaction to give aldehyde **5**.

Problem

1.
2.

5 (OTHP) → 6 ($SiMe_3$, OH)

Tips

$Me_3Si–CHBr_2$

12

- The first step is an olefination.
- The process leads to construction of a vinylsilane.
- Chromium(II) chloride is utilized in this reaction.
- Dibromide **12** is reduced with chromium(II) chloride.
- The THP protecting group can be cleaved with acid.

Solution

The first step is a *Takai* reaction.[4] Geminal dihalide **12** is presumably converted in the presence of chromium(II) chloride into the dichromium complex **13**, which reacts with aldehyde **5** to give product **14**.

This method is well-suited to preparing *trans* double bonds from aldehydes and geminal dihalides. The *Takai* reaction leads to *E*/*Z* selectivities >90:10, often even 99:1 (for further details see Chapter 13).

The second reaction, cleavage of the THP protecting group, is conducted under acidic conditions with PTSA in methanol, producing allylic alcohol **6**. With few exceptions, attempts to prepare alkenylsilanes with *Wittig* reagents containing silyl groups have proven unsuccessful.[5]

1. $CrCl_2$, $CHBr_2TMS$ (**12**), THF, RT, 86%.
2. PTSA, MeOH, RT, 85%.

Me$_3$Si–CHBr$_2$ **12** → Cr(II), THF → Me$_3$Si–CH(Cr(III)–Br)$_2$ **13** → + **5** → **14** (OTHP, SiMe$_3$)

Discussion

Apart from the *Takai* method and titanium reagents such as **15**, silyl reagents **16** and **17** frequently find application in the synthesis of vinylic silanes from carbonyl compounds. Reagent **16** can be utilized with aldehydes and non-enolizable ketones in a reaction analogous to the *Peterson* olefination. Reagent **17** also reacts successfully with enolizable ketones.[6]

CpTi(CH$_2$CH$_2$SiMe$_3$)$_3$ **15**

Me$_3$Si–CH$_2$–SiMe$_2$R

16: R = Me
17: R = OMe

Problem

6 (OH, SiMe$_3$) → 1. 2. → **7** (PhSO$_2$, SO$_2$Ph, SiMe$_3$)

Tips

- In this reaction sequence, allylic alcohol **6** is converted into **7** with a double inversion of configuration, and thus retention.
- In the first step the hydroxy group is transformed into a good leaving group, which is cleaved in the second step as CO_2.
- The first reaction involves methyl chloroformate (**18**) and a base.
- The allylic carbonate is substituted nucleophilically by a methylene bissulfone.
- The process occurs with transition-metal catalysis.
- The transition metal in question is palladium.

MeO–C(=O)–C **18**

Solution

Allylic alcohol **6** is first reacted with acid chloride **18** in the presence of pyridine to give allylic carbonate **19**. An allylic alkylation of sodium salt **20**, catalyzed by palladium in the oxidation state 0, leads then to compound **7**.

Allylic carbonate **19** adds oxidatively with loss of CO_2 and inversion of configuration to a palladium(0) complex, which produces allylpalladium species **21**. The σ complex **21a** of the latter is in equilibrium with the corresponding π complex **21b**.[7] After cleavage of methoxide, this allylpalladium species (**22**) undergoes nucleophilic attack regioselectively and again with inversion (and thus overall retention of configuration) by metallated bissulfone[8] **20**. The result is vinylsilane **7** substituted in the allylic position, together with the palladium catalyst.[9]

The regiochemistry of this allylic substitution is determined primarily by steric factors.[9] Substitution occurs from the less hindered side of allylic complex **22**. This behavior is typical for attack by soft nucleophiles. Soft nucleophiles are distinguished by the fact that their charge can be stabilized by resonance. Examples include not only sulfones but also nitriles, nitro compounds, ketones, and esters of carboxylic acids.

Hard nucleophiles such as alkyllithium and alkylmagnesium compounds transfer their alkyl residues initially via a transmetallation reaction to the palladium atom in **22**. Reaction product **24** is then released from **23** through a reductive elimination. Because this reductive elimination occurs with retention of configuration, the result here is an overall inversion of configuration. Nevertheless, reactions of this type have little preparative significance.

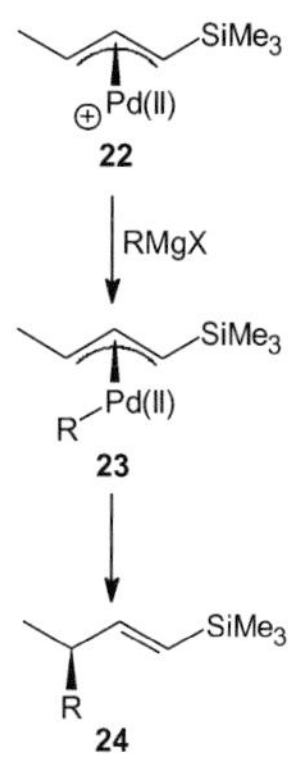

SiMe3 ⊕Pd(II) **22** RMgX SiMe3 R Pd(II) **23** SiMe3 R **24**

In order to assure the regiochemistry of the allylic alkylation employed here it is necessary to introduce the trimethylsilyl group into starting material **6** as a "dummy substituent."

1. **18**, pyridine, CH_2Cl_2, 0 °C → RT.
2. $Pd_2(dba)_3$, 1,2-bis(diphenylphosphino)ethane, **20**, THF, RT, 80% over two steps.

Problem

SiMe3 PhO2S SO2Ph **7** → PhO2S SO2Ph **8**

Tips

- The reaction is a protodesilylation.
- A sulfonic acid is added.

Solution

The trimethylsilyl group is exchanged for a hydrogen atom using PTSA in acetonitrile, a reaction referred to as a "protodesilylation." Protonation of vinylsilane **7** leads in the process first to cation **25**, which is stabilized by the β-effect of silicon.[10] A nucleophile subsequently abstracts the TMS group, producing olefin **8**. Vinylsilanes can be transformed into olefins not only with acid,[10] but also with fluoride[10,11] or with catalytic amounts of iodine in the presence of water.

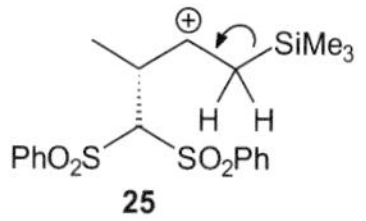

⊕ SiMe3 H H PhO2S SO2Ph **25**

PTSA, CH_3CN, reflux, 96%.

Problem

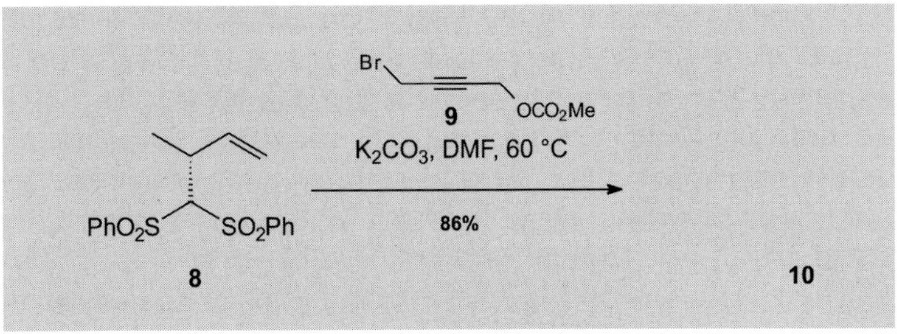

Tips

- Bissulfone **8** is converted into the propargylic compound **9** with chain extension.
- The bissulfone group is a CH acid.

Solution

Starting material **8** is treated with the propargylic bromide **9** in the presence of potassium carbonate as base in a nucleophilic substitution, producing compound **10**.

Problem

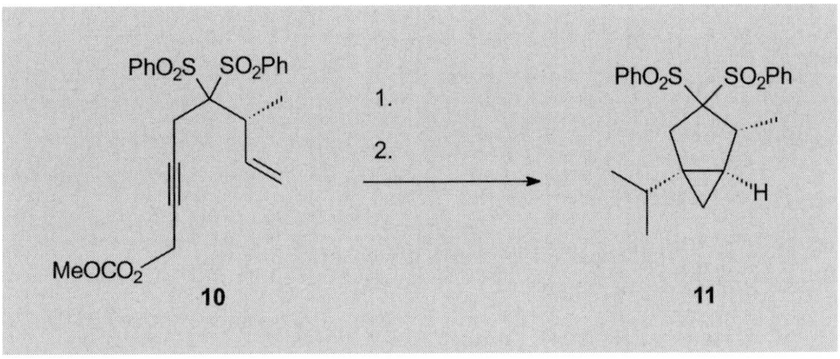

Tips

- The first reaction accomplishes cyclization to the bicyclic system.
- Palladium in the oxidation state 0 induces a cascade of successive reactions starting from **10**.
- Analogously to allylic carbonate **19**, propargylic carbonate **10** also reacts in the first step with a palladium complex.
- The propargylic palladium compound is in equilibrium with an allenic palladium compound.
- Starting from this species, two successive cyclizations occur leading to a vinylpalladium species.
- One carbon atom is still missing.
- A transmetallation takes place from zinc to palladium in the context of a cross-coupling reaction.

- The product of the first reaction contains a terminal double bond, which is hydrogenated in the second reaction to an isopropyl group.

Solution

The reaction employed here is a palladium-catalyzed domino 1,6-enyne cyclization.[12] Carbonate **10** adds oxidatively to a complex containing palladium in the zero oxidation state.[13] The resulting intermediate, **26a**, is in equilibrium with the palladium-allenyl species **26b**.[14]

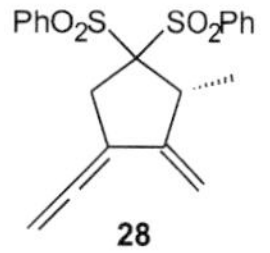

The terminal double bond is inserted into the palladium–carbon bond of **26b** with formation of alkylpalladium compound **27**. This substance has available to it two fundamental reaction pathways: it might suffer β-hydride elimination to product **28** with an exocyclic double bond on the cyclopentane ring system, or a further carbopalladination of the allenic system might occur leading to construction of the anellated cyclopropane ring in **29**.

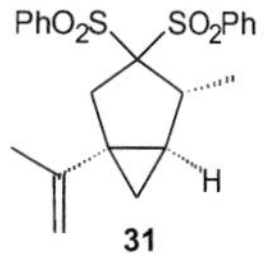

In fact, the sole product is **29**, because in this case β-hydride elimination occurs more slowly than the competing second carbopalladination. Vinylpalladium species **29** is converted via **30** into product **31** through a cross-coupling reaction with dimethylzinc. Subsequent hydrogenation of the terminal double bound of **31** with platinum dioxide in acetic acid/ethyl acetate in a second reaction results in product **11**.

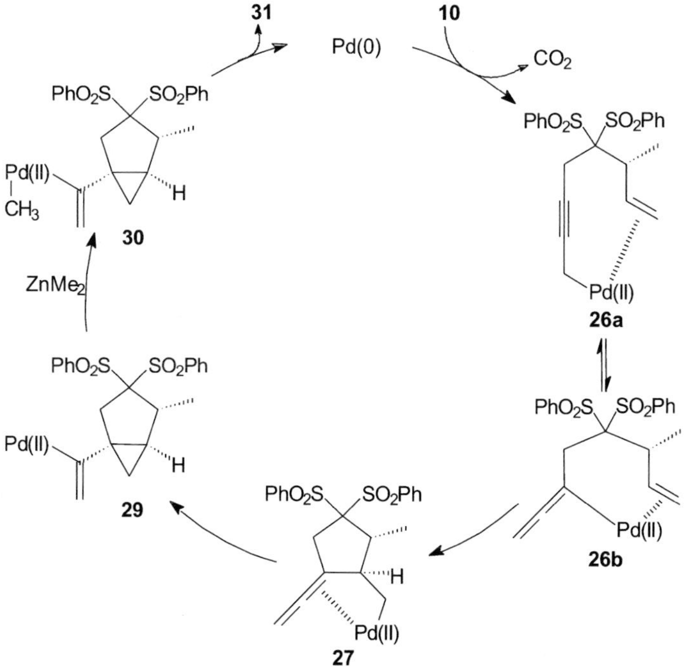

PhO_2S SO_2Ph
28

PhO_2S SO_2Ph H
31

31 10 Pd(0) CO_2
PhO_2S SO_2Ph Pd(II) CH_3 H **30**
PhO_2S SO_2Ph Pd(II) **26a**
$ZnMe_2$
PhO_2S SO_2Ph Pd(II) H **29**
PhO_2S SO_2Ph Pd(II) **26b**
PhO_2S SO_2Ph H Pd(II) **27**

1. $Pd_2(dba)_3$, $ZnMe_2$, Et_2O, P(2-furyl)$_3$, reflux, 91%.
2. H_2, PtO_2, HOAc, EtOAc, 50 °C, 96%.

Problem

PhO_2S SO_2Ph H **11** → 1. 2. → O H **1** **(-)-α-thujone**

Tips

- In the first step, a sulfone group is reductively removed with the aid of a reducing agent.
- The reducing agent is an amalgam. Mercury is alloyed with an element from the third main group.
- Base is used in a second step to generate a carbanion α to the sulfonyl group.
- This carbanion is oxidized by a molybdenum complex.

Solution

SO_2Ph H **32** → Li SO_2Ph H **33** → (MOPH) MoO_5 · pyridine · HMPA → $MO_4L_2^{\ominus}$ O SO_2Ph H **34**

Bissulfone **11** is reduced with aluminum amalgam to sulfone **32**. Epimers are obtained at this stage, which are transformed into the common product **1** through subsequent oxidation to a ketone.

In the course of oxidation of the sulfonyl group in **32**, lithium diisopropylamide is used to form the α-sulfonyl carbanion **33**. This anion reacts with MOPH via complex **34** to give natural product **1**. The MOPH complex utilized here is crystalline and stable to air over brief periods of time, so it is relatively convenient to employ. It has become a standard reagent for α-hydroxylation of CH-acidic compounds.[15]

1. Al/Hg, THF/H_2O (19 : 1), RT.
2. LDA, THF, –78 °C; MOPH, 60% over two steps.

8.4 Summary

The preceding 12-step sequence was the first published enantioselective synthesis of (–)-α-thujone (**1**). With the aid of gas chromatography over a chiral capillary column it was established that the enantiomeric excess (*ee*) in the resulting natural product **1** was >99.2%.

Chiral information was introduced through the C3-building block **5**, derived from lactic acid. The remaining two stereogenic centers in the natural product were provided under substrate control in the course of a domino cyclization.

A trimethylsilyl group in compound **6** serves as a dummy substituent in order to control the regiochemistry of the allylic alkylation reaction. *Oppolzer* added the necessary vinylsilane structural element with a *Takai* reaction.

The key step in this synthesis is the palladium-catalyzed domino-1,6-enyne cyclization, which creates the bicyclic skeleton of the natural product in a single diastereoselective step.

O
OTHP
5

SiMe$_3$
OH
6

8.5 References

1 V. Hach, *J. Org. Chem.* **1977**, *42*, 1616.

2 R. Grigg, R. Rasul, J. Redpath, D. Wilson, *Tetrahedron Lett.* **1996**, *37*, 4609; W. Oppolzer, A. Pimm, B. Stammen, W.E. Hume, *Helv. Chim. Acta* **1997**, *80*, 623.

3 S.P. Acharya, H.C. Brown, A. Suzuki, S. Nozawa, M. Itoh, *J. Org. Chem.* **1969**, *34*, 3015.

4 K. Takai, Y. Kataoka, T. Okazoe, K. Utimoto, *Tetrahedron Lett.* **1987**, *28*, 1443; T. Okazoe, K. Takai, K. Utimoto, *J. Am. Chem. Soc.* **1987**, *109*, 951.

5 H. Gilman, R.A. Tomasi, *J. Org. Chem.* **1962**, *27*, 3647; F. Plénat, *Tetrahedron Lett.* **1981**, *22*, 4705.

6 H. Sakurai, K. Nishiwaki, M. Kira, *Tetrahedron Lett.* **1973**, *14*, 4193; B.T. Gröbel, D. Seebach, *Chem. Ber.* **1977**, *110*, 852; T.F. Bates, R.D. Thomas, *J. Org. Chem.* **1989**, *54*, 1784; N.A. Petasis, I. Akritopoulou, *Synlett* **1992**, 665.

7 J. Tsuji, M. Yuhara, M. Minato, H. Yamada, F. Sato, Y. Kobayashi, *Tetrahedron Lett.* **1988**, *29*, 343; J. Tsuji, *Tetrahedron* **1986**, *42*, 4361.

8 A. Herve du Penhoat, M. Julia, *Tetrahedron* **1986**, *42*, 4807.

9 L.S. Hegedus, *Transition metals in the synthesis of complex organic molecules*, Wiley-VCH, Weinheim 1995; B.M. Trost, D.L. Van Vranken, *Chem. Rev.* **1996**, *96*, 395.

10 D.P. Mathews, R.S. Gross, J.R. McCarthy, *Tetrahedron Lett.* **1994**, *35*, 1027; S.E. Gibson, *The roles of boron and silicon*, Wiley-VCH, Weinheim 1995.

11 R. Angell, P.J. Parsons, A. Naylor, *Synlett* **1993**, *3*, 189.

12 H.G. Schmalz, *Nachr. Chem. Tech. Lab.* **1994**, *42*, 270; W. Cabri, I. Candiani, *Acc. Chem. Res.* **1995**, *28*, 2; L.E. Overman, D.J. Poon, *Angew. Chem. Int. Ed. Engl.* **1997**, *36,* 518; *Angew. Chem. Int. Ed. Engl.* **1997**, *36*, 518.

13 J. Tsuji, T. Mandai, *Angew. Chem. Int. Ed. Engl.* **1995**, *34,* 2589; *Angew. Chem. Int. Ed. Engl.* **1995**, *34*, 2589; M. Ahmar, B. Cazes, J. Gore, *J. Chem. Soc., Chem. Commun.* **1984**, *25*, 4505.

14 S. Ma, E.-I. Negishi, *J. Am. Chem. Soc.* **1995**, *117*, 6345; R.C. Larock, N.G. Berrios-Pea, C.A. Fried, *J. Org. Chem.* **1991**, *56*, 2615.

15 E. Vedejs, D.A. Engler, J.E. Telschow, *J. Org. Chem.* **1978**, *43*, 188.

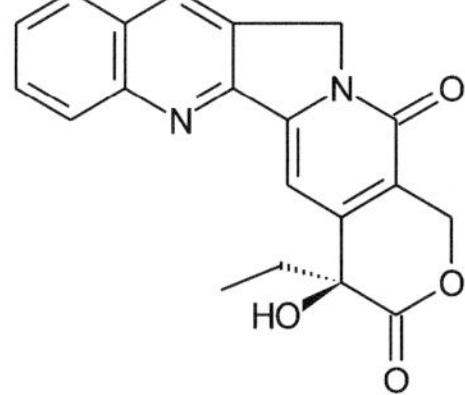

9

(+)-Camptothecin: Ciufolini (1996)

9.1 Introduction

Camptothecin (**1**) was first extracted from the heartwood of the tree species *Camptotheca Acuminata Nyssaceae* by *Wall*[1] in 1966. Its structure was verified by X-ray analysis[2] of its C-20 iodoacetate ester.

The pentacyclic quinoline alkaloid camptothecin (**1**) contains in addition to the quinoline system (rings A and B) a pyridone unit (ring D) and a lactone ring (E). Compound **1** also includes one stereogenic center (at C-20), which has the *S*-configuration.

Camptothecin (**1**) is a highly effective antitumor agent.[3] It is assumed that the substance interferes with the process of unwinding the DNA helix through inhibition of topoisomerase I.[4] Numerous structure–activity investigations have made it possible to identify analogues of camptothecin with similar antitumor activity and especially with increased water solubility. Variation is possible at C-7 and at C-9 through C-11 without loss of activity, whereas substitution at other positions reduces the activity or even eliminates it altogether.[5]

The first total synthesis of racemic camptothecin was reported by *Stork* and *Schultz*[6] as early as 1971. A number of total syntheses followed, many of which were also of racemic material.[7] Enantiomerically pure natural product has been prepared by the method of kinetic resolution,[8] through use of chiral auxiliaries,[9] and by *Sharpless* dihydroxylation,[10] an example of a catalytic asymmetric route.

Ciufolini's[11] retrosynthetic analysis of natural product **1** leads not only to a quinoline derivative but also to aldehyde **2**, which contains the stereogenic center at C-20, as well as to cyanoacetamide (**3**), which provides carbon atoms C-16a, C-16, and C-17 in camptothecin (**1**).

9.2 Overview

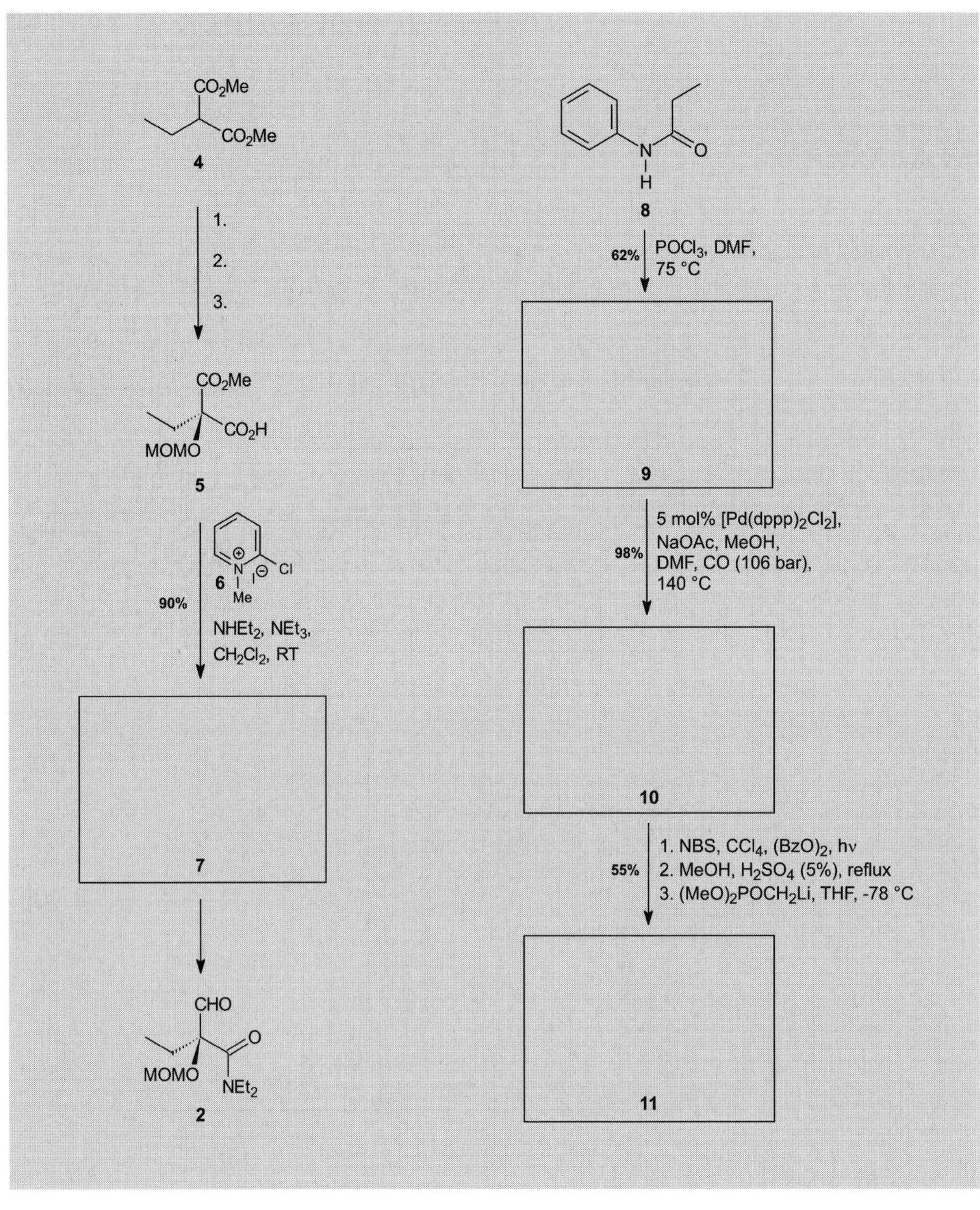
CO2Me
CO2Me
4
1.
2.
3.
CO2Me
CO2H
MOMO
5
6
Cl
Me
90%
NHEt2, NEt3,
CH2Cl2, RT
7
CHO
O
MOMO
NEt2
2
N
O
H
8
62%
POCl3, DMF,
75 °C
9
5 mol% [Pd(dppp)2Cl2],
NaOAc, MeOH,
98%
DMF, CO (106 bar),
140 °C
10
1. NBS, CCl4, (BzO)2, hν
55%
2. MeOH, H2SO4 (5%), reflux
3. (MeO)2POCH2Li, THF, -78 °C
11

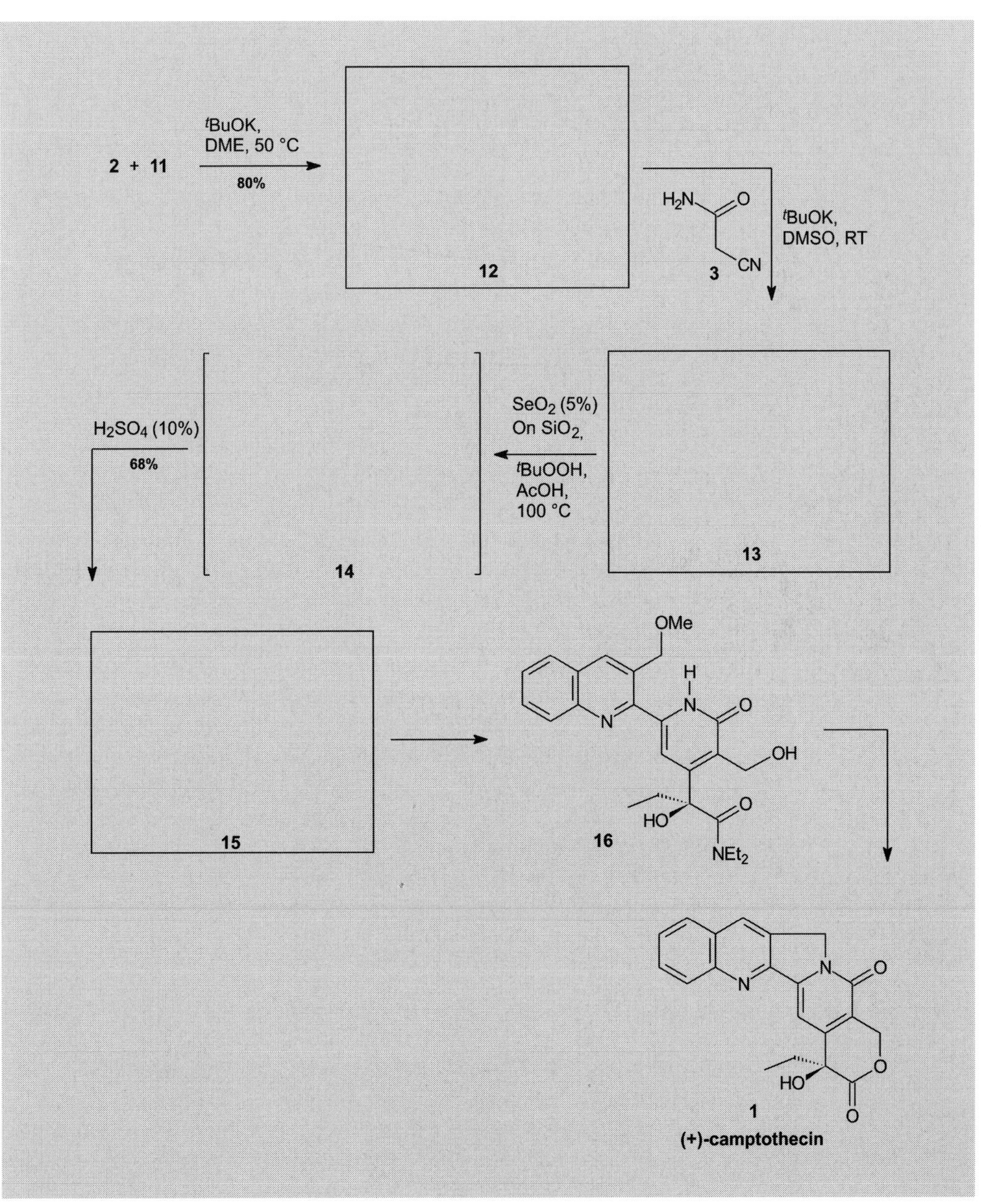

2 + 11
ᵗBuOK, DME, 50 °C
80%
12
H2N
O
3
CN
ᵗBuOK, DMSO, RT
13
SeO2 (5%) On SiO2,
ᵗBuOOH, AcOH, 100 °C
14
H2SO4 (10%)
68%
15
OMe
H
N
O
N
OH
HO
O
NEt2
16
N
O
N
O
HO
O
1
(+)-camptothecin

9.3 Synthesis

Problem

Tips

- This reaction sequence transforms an achiral substrate, *meso* compound **4**, into the chiral substance **5**.
- An additional functionality is introduced in the first step.
- The process in question is a hydroxylation.
- The reagent utilized here contains three oxygen atoms.
- The hydroxy group is protected in the second step.
- The third step occurs enantioselectively.
- This reaction is a hydrolysis.
- Which of the following reagents are suited to accomplishing enantioselective hydrolyses: a) H^+, H_2O; b) OH^-, H_2O; c) enzymes?

Solution

The dimethyl ester of 2-ethylmalonic acid (**4**) is first adsorbed onto silica gel and then hydroxylated with ozone. The resulting hydroxy group is protected in a second step in the form of a methoxymethyl (MOM) ether.[12]

Enantioselective enzymatic hydrolysis[13] with pig liver esterase (PLE) transforms the *meso* substrate into chiral compound **5** with >98% *ee*. This enzyme is capable of differentiating between the two enantiotopic ester groups on the prochiral carbon atom and hydrolyzing only one of them to a carboxylic acid. Maximum enantioselectivity is achieved by carrying out the reaction in 25% aqueous DMSO solution at 35 °C.

1. SiO_2, O_3, RT.
2. 2. MOMCl, NEt^iPr_2, CH_2Cl_2, RT, 100%.
3. PLE, DMSO, H_2O, pH 6.8–7.4, 35 °C, 90%, >98% *ee*.

Discussion

Enzymes catalyze a broad spectrum of reactions. They often require such coenzymes as the nicotinamides NADPH and NADH or a nucleoside triphosphate like ATP together with cofactors, usually metal ions. Hydrolases, including PLE, are exceptions in this regard. They complete their tasks without the need for coenzymes. Enzyme-catalyzed asymmetric syntheses can be conducted either with cell-free enzymes or with microbial systems (i.e., enzymes included within cells).[14]

Enzymes are grouped into six classes: 1. Oxidoreductases – enzymes that catalyze oxidation and reduction reactions. An example is the baker's yeast microbial system,[15] which catalyzes the stereospecific reduction of carbonyl groups to alcohols; 2. Transferases – enzymes that catalyze the transfer of acyl, phosphate, glycosyl, carboxyl, and formyl groups; 3. Hydrolases – enzymes that catalyze the cleavage and formation of esters, amides, peptides, and glycosides, a category that includes the pig liver esterase utilized here;[16] they are preferentially employed for reactions of diesters of dicarboxylic acids, which can lead to chiral monoesters if one takes advantage of the meso trick; 4. Lyases – enzymes that catalyze additions to C-C, C-N, and C-O double bonds, as well as the corresponding reverse reactions; 5. Isomerases – enzymes that catalyze rearrangements and *E*/*Z* isomerization at double bonds; and 6. Ligases – enzymes that catalyze the coupling of a pair of molecules with the formation of a C-C, C-N, C-O, or C-S bond.

classes of enzyme:

1. oxidoreductases
2. transferases
3. hydrolases
4. lyases
5. isomerases
6. ligases

Enzymes display a number of characteristic features.[17] They catalyze enantioselective reactions, are usually substrate specific, usually display their greatest catalytic activity in aqueous systems, and are generated in nature in only one asymmetric form. The greatest advantage of enzyme-catalyzed asymmetric synthesis is that the availability of the correct enzyme makes it possible to prepare large amounts of an enantiomerically pure compound with relatively little effort.

Problem

CO_2Me
CO_2H
MOMO
5
6 N Cl Me
$NHEt_2$, NEt_3,
CH_2Cl_2, RT
90%
7

Tips

- Reagent **6** is in a position to activate the reactive center.
- The carboxyl group reacts first with the *Mukaiyama* reagent **6**.
- Diethylamine acts as a nucleophile.
- Which of the following is obtained as the product: a) an acid; b) an ester; c) an amine; d) an aldehyde; e) an amide; or f) an alcohol?

Solution

The *Mukaiyama* reagent,[18] *N*-methyl-2-chloropyridinium iodide (**6**), transforms carboxylic acid **5** into the amide **7**. The acid is first activated in situ in the form of pyridinium salt **17** by an S_N reaction with the *Mukaiyama* reagent (**6**). This activation is a result of preventing resonance stabilization of the C-O double bond in the positively charged aryl ester **17**.

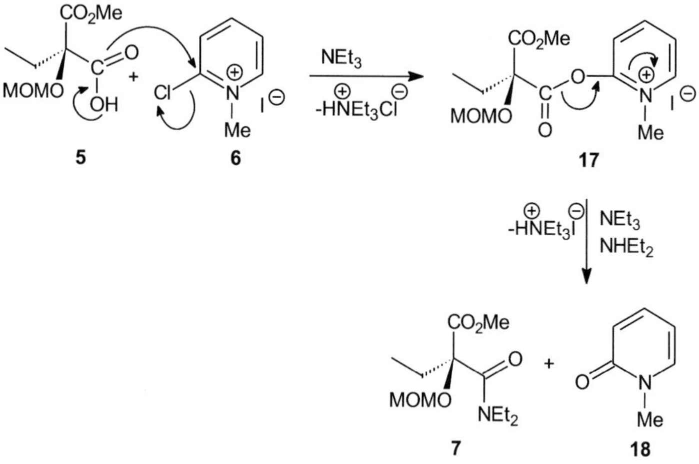

Activated species **17** reacts with diethylamine as a nucleophile to produce amide **7** together with *N*-methylpyridone 18.

Discussion

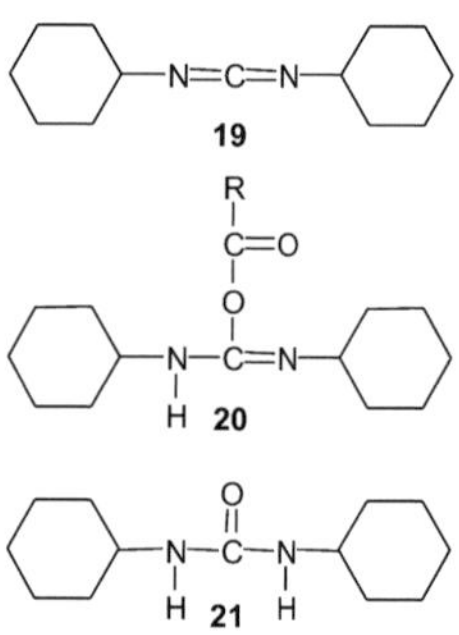

Another in situ procedure for activating carboxylic acids utilizes carbodiimides, such as dicyclohexylcarbodiimide (DCC). DCC (**19**) plays an important role in peptide synthesis. Addition of a carboxylic acid to the C-N double bond leads to the activated species, an acyl isourea **20**, which upon attack by a nucleophile (and alcohol or an amine) releases the corresponding ester or amide along with **21** (for the mechanism, see Chapter 5). However, in the conversion of **5** to **7** the DCC procedure gives poor results.

The *Yamaguchi* procedure constitutes another in-situ approach to the activation of carboxylic acids. Here the activated species is a mixed anhydride formed with trichlorobenzoic acid chloride (see Chapter 6).

Problem

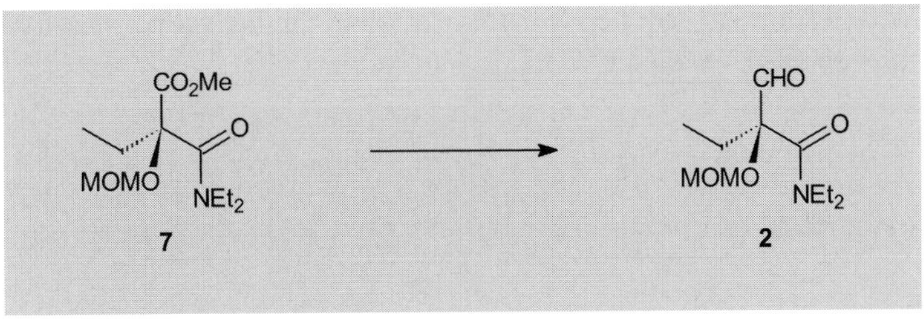

Tips

- The methyl ester is converted chemoselectively into an aldehyde. Which of the following reaction types is involved: a) oxidation; b) elimination; c) reduction; or d) hydrolysis?
- The reduction occurs as a single step.

Solution

Chemoselective reduction of methyl ester **7** to aldehyde **2** is possible with DIBAH. The metallated hemiacetal that results from addition of DIBAH to the carbonyl group of an ester usually decomposes rapidly in polar solvents like THF to an intermediate aldehyde. This then competes with the ester and, as a result of its higher electrophilicity, is reduced by DIBAH to an alcohol. However, ester **7** bears a methoxymethyl residue in its α-position, which stabilizes the metallated hemiacetal by chelate formation. Chelate complex **22** is protolytically cleaved by way of the hemiacetal only in the course of aqueous workup, so in this case the DIBAH reaction produces only aldehyde **2**, not the alcohol (see also Chapter 3). DIBAH, THF, –78 °C, 100%.

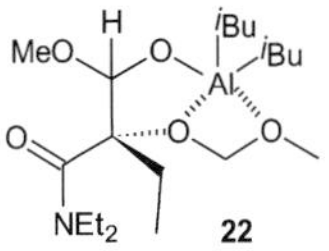

Discussion

The enantiomer *ent*-**2** is also easily accessible from **5** by methyl ester/diethylamide exchange and subsequent esterification with diazomethane to **23** followed by DIBAH reduction.[11]

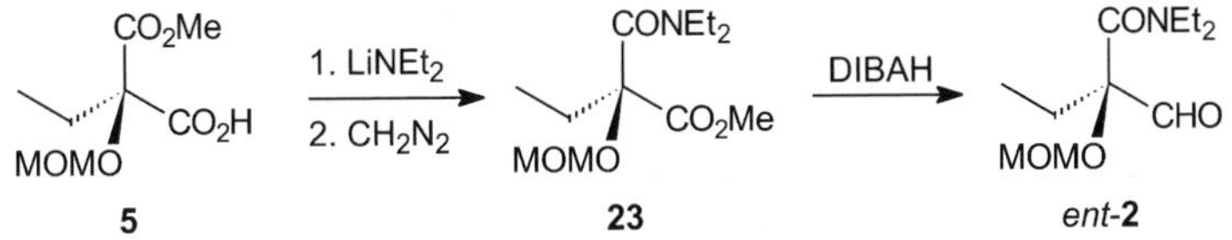

This synthetic sequence opens the possibility of obtaining both enantiomeric forms of aldehyde **2** with the aid of the enzyme. Enzymes differ from many chiral catalysts in that they are available in only a single enantiomeric form and therefore can lead to synthesis of only one enantiomer.

From *ent*-**2** the unnatural camptothecin *ent*-**1** was also obtained, with which it was possible to verify the absolute configuration of **2**.

What follows is synthesis of the quinoline derivative **11**, which is then coupled with aldehyde **2**.

Problem

POCl$_3$, DMF, 75 °C

62%

8 9

Tips

- A ring closure takes place.
- Anilide **8** reacts first with phosphoryl chloride and then with the product from reaction of DMF with phosphoryl chloride.
- The result is a quinoline.

Solution

Anilide **8** is transformed under *Vilsmeier* conditions into the quinoline derivative[19] **9**. In the process, phosphoryl chloride attacks the carbonyl group of anilide **8** and initially forms imine **24**, which is in equilibrium with the corresponding enamine **25**.

POCl$_3$

8 24 25

-H$^{\oplus}$ -HCl 27

-NHMe$_2$

9 29 28

26 -OPOCl$_2$ 27

The resulting enamine **25** then reacts with the *Vilsmeier* reagent **27**, which results from DMF and phosphoryl chloride by way of intermediate **26**. Elimination of hydrogen chloride leads to compound **28**. Ring closure presumably occurs via intermediate **29**. Rearomatization with elimination of dimethylamine produces the desired quinoline **9**.

Problem

5 mol% $[Pd(dppp)_2Cl_2]$,
NaOAc, MeOH,
DMF, CO (106 bar),
140 °C

98%

9 **10**

Tips

- Product **10** no longer contains a chlorine substituent.
- A methyl ester group is introduced, assembled from carbon monoxide and methanol.
- The first step is an oxidative addition of **9** to a palladium species, followed by CO insertion and finally a nucleophilic cleavage by methanol.

Solution

Quinoline derivative **9** is carbomethoxylated under palladium catalysis with 1,3-bis(diphenylphosphino)propane (dppp) as ligand to give **10**. Although aryl chlorides are usually unreactive in such carbonylation reactions, 2-haloquinolines have been found to be highly reactive.[20]

Initially, oxidative addition of the 2-chloroquinoline **9** to a Pd^0 species leads to complex **30**, which contains Pd^{II}. Carbonylation then occurs through CO insertion into the palladium–aryl bond, giving intermediate **31**. Finally, methanolysis releases methyl ester **10**, and the Pd^0 species is regenerated.

30

31

Problem

1. NBS, CCl_4, $(BzO)_2$,hv
2. MeOH, H_2SO_4, reflux
3. $(MeO)_2POCH_2Li$, THF, -78 °C

55%

10 **11**

Tips

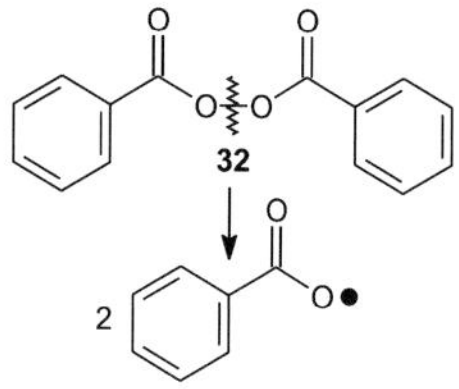

- Quinoline **10** is extended by one C_1 unit.
- The first step is a bromination reaction.
- Benzoyl peroxide (**32**) serves as a radical initiator.
- Where does radical attack occur: a) on the side chain, or b) in the ring?
- The second step is an S_N reaction.
- The functionality introduced in the first step undergoes methanolysis.
- The third step leads to a β-ketophosphonate.

Solution

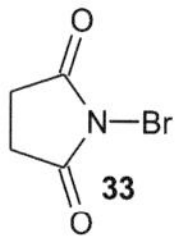

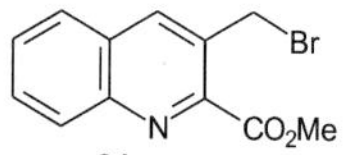

The first step constitutes a chemoselective radical bromination of the methyl group in quinoline **10** using *N*-bromosuccinimide (**33**), leading to compound **34**. Here benzoyl peroxide (**32**) acts as the radical initiator. A rule of thumb for chemoselectivity states that heat and light produce side-chain halogenation, whereas cold and catalysis favor halogenation of the aromatic nucleus.

In the second step, the bromine atom is displaced in an S_N reaction, generating compound **35**.

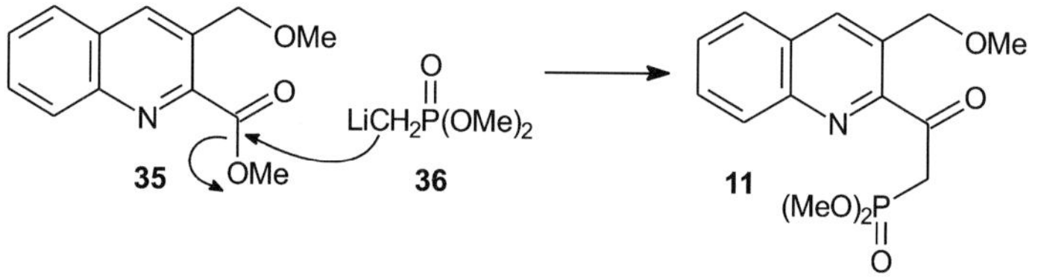

Acylation of lithiated phosphonic ester **36** with methyl ester **35** gives, through a *Corey–Kwiatkowski* reaction,[21] the β-ketophosphonic ester **11**.

Problem

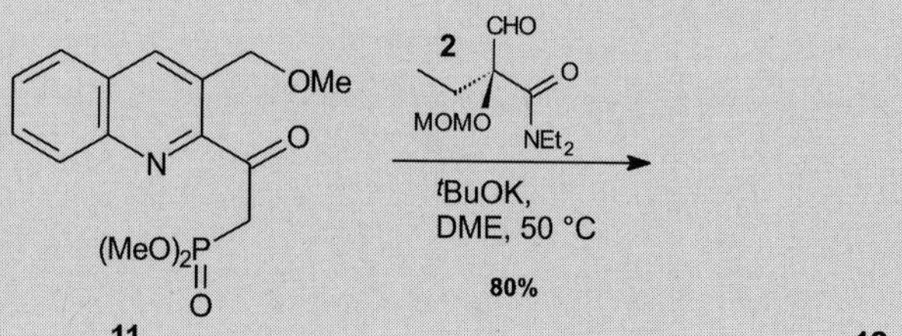

Tips

- An anion of compound **11** reacts with the most reactive site in molecule **2**. Which is this: a) the MOM-protected hydroxy group; b) the amide function; or c) the aldehyde?
- Phosphorus atoms show a great affinity for oxygen.
- A phosphoric ester is cleaved.
- The overall reaction is an olefination.

Solution

Condensation occurs between the potassium salt of β-ketophosphonic ester **11** and aldehyde **2** to produce the C-C double bond in α,β-unsaturated ketone **12**. This reaction is known as the *Horner-Wadsworth-Emmons* procedure.[22] The phosphoric ester anion **39** is formed as a byproduct. Such *Horner-Wadsworth-Emmons* reactions generally occur with *trans* selectivity. One possible mechanism involves alkoxide **37**, which cyclizes to oxaphosphetane **38**. This in turn cleaves to give enone **12** and the phosphoric ester anion **39**. However, this mechanism has not yet been definitively proven.

OMe O MOMO O NEt2 12

(MeO)2P—CH—C—Ar R—C—H 11 2 → P(OMe)2 R Ar 37 → O—P(OMe)2 R Ar 38 → R Ar + O=P(OMe)2 12 39

Discussion

The term *Horner-Wadsworth-Emmons* reaction refers not only to the illustrated conversion of β-ketophosphonic esters **11** into α,β-ketones **12**, but also to the reaction of β-alkoxycarbonylphosphonic esters **40** to give α,β-unsaturated esters.

Whereas β-keto compounds are accessible as described through metallated phosphonic acid esters (the *Corey-Kwiatkowski* reaction), β-alkoxycarbonyl compounds **40** as starting materials for the *Horner-Wadsworth-Emmons* reaction are obtained via the *Arbusov* reaction, an S_N2 process involving phosphorus nucleophiles.

$P(OMe)_3 + Br—CH_2—CO_2R$ → (-MeBr) $(MeO)_2P(O)—CH_2—CO_2R$ **40**

$(CF_3-CH_2-O)_2P(=O)-CH_2-CO_2R$ **41**

The *trans* selectivity in the *Horner-Wadsworth-Emmons* reaction can be reversed by structural variation in the phosphonic ester moiety, the so-called *Still-Gennari* variant. In this case trifluoro-substituted ethoxy residues are introduced in the form of phosphonic ester **41**, providing access to *cis*-substituted acrylic esters.

Problem

OMe, O, N, MOMO, O, NEt_2, **12**; H_2N, O, **3**, CN; ^{t}BuOK, DMSO, RT; **13**

Tips

- Cyanoacetamide **3** contains an acidic CH group that can easily be deprotonated.
- The α,β-unsaturated carbonyl function in compound **12** reacts with the nucleophile.

Solution

Vinylogous ketone **12** reacts with cyanoacetamide (**3**) in a *Michael* reaction to give a diastereomeric mixture of chain and ring tautomers **13a** and **13b**.

OMe, O, N, **12**, MOMO, O, NEt_2; H_2N, O, CN, **3**; $H^{\oplus}$

OMe, OH, H, N, O, CN, **13b**, MOMO, O, NEt_2 ⇌ OMe, O, $CONH_2$, CN, **13a**, MOMO, O, NEt_2

Problem

Tips

- Acetic acid functions as a proton source and accelerates the elimination of water following a cyclization.
- What occurs next is an aromatization.
- The product is a 2-hydroxypyridine, which is in equilibrium with its tautomer, pyridone **14**.
- Selenium dioxide serves as an oxidizing agent.

Solution

The process begins with complete ring closure to compound **13b**, since acid-catalyzed elimination of water to the dihydropyridone shifts the equilibrium between compounds **13a** and **13b**.

Selenium dioxide as a dehydrogenation reagent[23] is capable of aromatizing the resulting ring to a 2-hydroxypyridine, which is present largely in its tautomeric form as pyridone **14**. This product is not isolated, but is instead immediately reacted further.

Discussion

3-Cyano-2-pyridones **44** are also available through a one-step synthesis starting with an enone **42** and cyanoacetamide (**3**). In this case under an oxygen atmosphere the results is an in-situ oxidation of the adduct **43**.[24]

In the course of the reaction, *Michael* adduct **43** cyclizes initially under base catalysis to the dihydropyridone **45**, which forms dianion **46**. Electron transfer (an SET process) leads to radical anion **47**, which is finally transformed into pyridone **44** through an aromatization that includes hydrogen transfer and another SET process.

For unknown reasons this one-step synthesis was not successful with enone **12**, so that in this case a separate oxidation with selenium dioxide was required subsequent to the *Michael* addition.

Problem

Tips

- What becomes of cyanide groups in the presence of strong aqueous acid: a) reduction; b) hydrolysis; or c) no reaction?
- The MOM protecting group is not stable to acid.
- A cyclization occurs.
- The product is a lactone.

Solution

Treatment with aqueous sulfuric acid cleaves the MOM protective group (which consists of an acetal structure) to give compound **48**.

The newly released hydroxy group adds to the nitrile function with formation of a cyclic imidoester (**49**), which then hydrolyzes to lactone **15** via intermediate **50** in a yield of 68% starting from compound **12**.

R =

14 48 49 50 15

-NH_3

Problem

15 16

Tip

- The lactone is reductively cleaved to diol **16**.

Solution

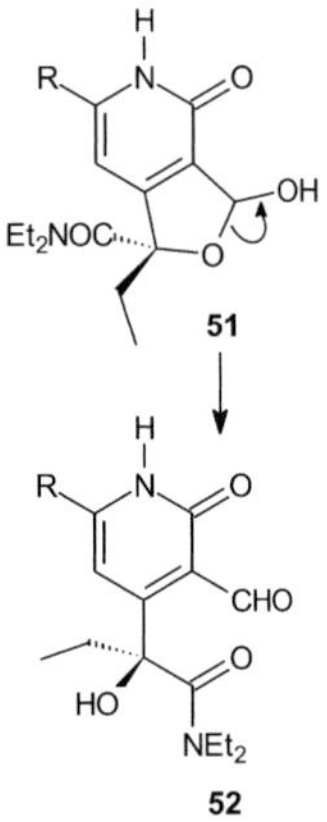

Reduction to diol **16** was achieved with a variation of the method of *Luche*,[25] involving sodium borohydride and cerium trichloride. The initial product is cyclic hemiacetal **51**, which is opened and reduced via intermediate **52** ultimately to diol **16**.

Numerous reducing agents were tried at this point unsuccessfully. For example, lithium aluminum hydride destroyed the substrate, whereas DIBAH or lithium borohydride in THF and sodium borohydride in ethanol led to reduction of the quinoline system. On the other hand, both potassium borohydride (either with or without 18-crown-6) and zinc borohydride (with or without ethanol) produced no reaction at all. Lithium triethylborohydride resulted in demethoxylation, and sodium borohydride in refluxing THF gave a 45% yield of diol **16** together with overreduced product.

10 eq. $NaBH_4$, 2.5 eq. $CeCl_3$, EtOH, 0 °C → 45 °C, 95%.

Discussion

The *Luche* procedure is noted for its ability to reduce a keto group chemoselectively in the presence of an aldehyde. This can be explained by the fact that cerium trichloride selectively complexes with the ketone, causing its electrophilicity to increase so much that it surpasses that of the aldehyde.

Ketones form more stable Lewis acid-base complexes with electrophilic metal salts than do aldehydes as a result of the increased basicity of the carbonyl oxygen atom thanks to the +I effect of the ketone's alkyl groups.

Problem

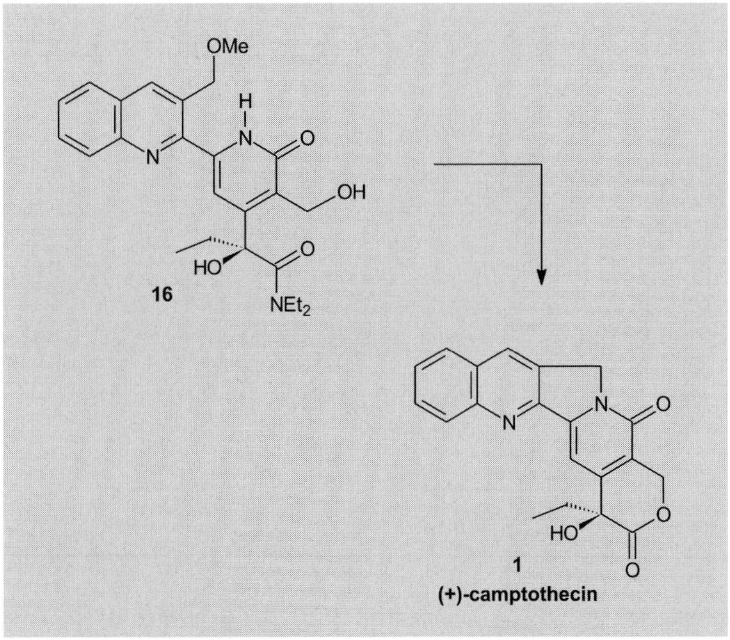

Tips

- Two rings must be closed to achieve the structure of target molecule **1**.
- A lactone ring is constructed.
- The same reagent must be capable of closing ring C as well.
- The transformation **14** → **15** was also a lactonization, but in this case ring C was not closed.

Solution

Ring E is closed by lactonization with 60% sulfuric acid and warming to 115 °C. Simultaneously, the methoxy group is activated (by protonation) with respect to nucleophilic attack by the pyridone nitrogen. Ring C is closed via an S_N reaction to produce the target molecule, (+)-camptothecin (**1**).

The reaction conditions here are more drastic than in the case of lactonization of **14** to **15** (H_2SO_4, 10%), for which reason it is only now that ring C is closed.
H_2SO_4 (60%), EtOH, 115 °C, 90%.

9.4 Summary

The key steps in the total synthesis above of (+)-camptothecin are the pyridone-forming ring closure of the D ring together with construction of the C ring through an N-4-C-5 coupling.

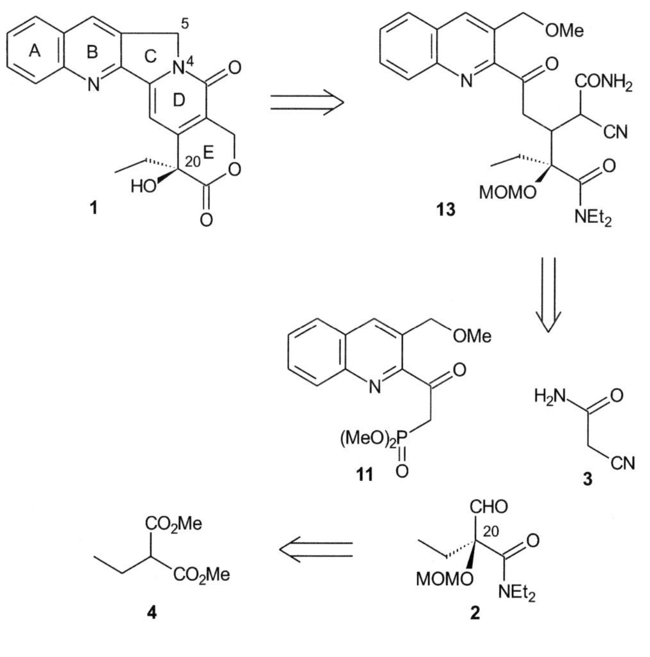

An appropriate precursor molecule **13** is accessible by condensation of the three components quinoline **11**, the aldehyde **2**, and cyanoacetamide (**3**). The stereogenic center at C-20 is introduced in the form of aldehyde **2** very early in the process by an enzyme-catalyzed asymmetric hydrolysis with pig liver esterase.

In this way, (+)-camptothecin (**1**) could be prepared from the 2-ethylmalonic ester **4** in ten steps with an overall yield of 30%. The synthesis is thus at least twice as effective as previously published enantioselective syntheses of this natural product.

9.5 References

1 M.E. Wall, M.C. Wani, C.E. Cook, K.H. Palmer, A.T. McPhail, G. A. Sim, *J. Am. Chem. Soc.* **1966**, *88*, 3888.

2 A.T. McPhail, G.A. Sim, *J. Chem. Soc.* **1968**, 923.

3 M. Potmesil, *Cancer Res.* **1994**, *54*, 1431.

4 Y.H. Hsiang, L.F. Liu, *Cancer Res.* **1988**, *48*, 1722; C.D. Lima, J.C. Wang, A. Mondragon, *Nature* **1994**, *367*, 138.

5 D.P. Curran, S.-B. Ko, H. Josien, *Angew. Chem. Int. Ed. Engl.* **1995**, *34,* 2683.

6 G. Stork, A.G. Schultz, *J. Am. Chem. Soc.* **1971**, *93*, 4074.

7 A.G. Schultz, *Chem. Rev.* **1973**, *73*, 385.

8 E.J. Corey, D.N. Crouse, J.E. Anderson, *J. Org. Chem.* **1975**, *40*, 2140; M.C. Wani, A.W. Nicholas, M.E. Wall, *J. Med. Chem.* **1987**, *30*, 2317.

9 A. Ejima, H. Terasawa, M. Sugimori, H. Tagawa, *J. Chem. Soc., Perkin Trans. 1* **1990**, 27; D.L. Comins, M.F. Baevsky, H. Hong, *J. Am. Chem. Soc.* **1992**, *114*, 10971.

10 F.G. Fang, S. Xie, M.W. Lowery, *J. Org. Chem.* **1994**, *59*, 6142; D.P. Curran, S.-B. Ko, H. Josien, *Angew. Chem. Int. Ed. Engl.* **1995**, *34,* 2683.

11 M.A. Ciufolini, F. Roschangar, *Angew. Chem. Int. Ed. Engl.* **1996**, *35,* 1692.

12 M. Luyten, S. Müller, B. Herzog, R. Keese, *Helv. Chim. Acta* **1987**, *70*, 1250.

13 E. Santaniello, P. Ferraboschi, P. Grisenti, A. Manzocchi, *Chem. Rev.* **1992**, *92*, 1071.

14 H.J. Gais, H. Hemmerle, *Chemie in unserer Zeit* **1990**, *5*, 239.

15 S. Servi, *Synthesis* **1990**, 1; R. Csuk, B.J. Glänzer, *Chem. Rev.* **1991**, *91*, 49.

16 M. Ohno, M. Otsuka in *Organic Reactions*, *Vol. 37*, (Ed.: S.A. Kende), Wiley, New York 1989, p. 1.

17 H.C. Wong, G.M. Whitesides, *Enzymes in Synthetic Organic Chemistry, Tetrahedron Org. Chem. Ser.*, *Vol. 12*, Oxford Press, Oxford 1994, p. 1.

18 T. Mukaiyama, M. Usui, E. Shimada, K. Saigo, *Chem. Lett.* **1975**, 1045.

19 O. Meth-Cohn, S. Rhouati, B. Tarnowski, A. Robinson, *J. Chem. Soc., Perkin Trans. 1* **1981**, 1537.

20 Y. Ben-David, M. Portnoy, D. Milstein, *J. Am. Chem. Soc.* **1989**, *111*, 8742.

21 E.J. Corey, G.T. Kwiatkowski, *J. Am. Chem. Soc.* **1966**, *88*, 5654.

22 T. Rein, O. Reiser, *Acta Chem. Scand.* **1996**, *50*, 369; R. Brückner, *Reaktionsmechanismen*, Spektrum Akademischer Verlag, Heidelberg 1996, p. 319.

23 B.R. Chmabra, K. Mayano, T. Ohtsuka, H. Shirahama, T. Matsumoto, *Chem. Lett.* **1981**, 1703.

24 R. Jain, F. Roschangar, M.A. Ciufolini, *Tetrahedron Lett.* **1995**, *36*, 3307.

25 J.L. Luche, L. Rodriguez-Hahn, P. Crabbé, *J. Chem. Soc., Chem. Commun.* **1978**, 601.

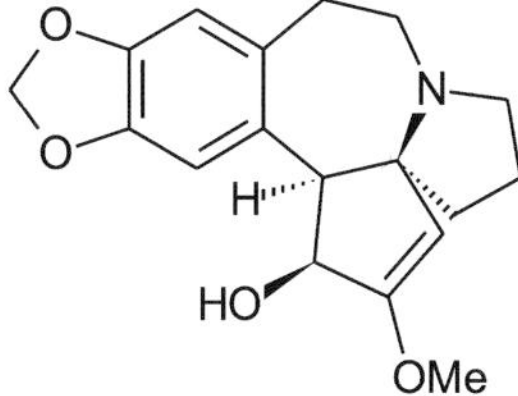

10

(–)-Cephalotaxin: Mori (1995)

10.1 Introduction

Yews of the genus *Cephalotaxus*, native to southeast Asia, contain a group of unusual pentacyclic alkaloids the parent structure of which is called cephalotaxin (**1**). The presence of alkaloids in the plants was first recognized by *Wall*[1] in 1954, and the compounds were first isolated by *Paudler*[2] in 1963. Their structure was determined in 1969 by *Abraham*[3] and the absolute configuration by *Bates* and *Powell*,[4] in both cases on the basis of X-ray crystallography.

Parry conducted experiments that demonstrated the biosynthetic origin of cephalotaxin (**1**) from one molecule each of tyrosine and phenylalanine, indicating that the compound should be regarded as a modified 1-phenethyltetrahydroisoquinoline alkaloid.[5]

Powell described as early as 1970 the antitumor activity of the related harringtonins **2**–**5** with respect to the mouse leukemias P-388 and L-1210.[6] Activity in terms of human cancers has also been established,[7] and is currently the subject of clinical studies.

Cephalotaxin has since become the goal of intensive synthetic efforts as a result not only of applications of the C-3 esters harringtonin (**2**) and homoharringtonin (**3**) in cancer therapy, but also because of the compound's unique structure, which incorporates a 1-azaspiro[4.4]nonane unit. The first two total syntheses were reported simultaneously by *Weinreb* and *Semmelhack* in 1972.[8] Other preparations of the alkaloid in racemic form followed[9] before *Mori*[10] succeeded in 1995 in providing the first totally synthetic approach to enantiomerically pure (–)-cephalotaxin (**1**).

R =

H **1**: cephalotaxin

2: harringtonin

3: homoharringtonin

4: deoxyharringtonin

5: isoharringtonin

10.2 Overview

^{t}BuCHO, TFA (cat.), pentane, water separator

74%

6

R-proline

7

LDA, THF, -78 °C;

8

-78 °C

85%

9

1.

2.

3.

4.

5.

10

1.

2.

11

1.

2.

12

14

13

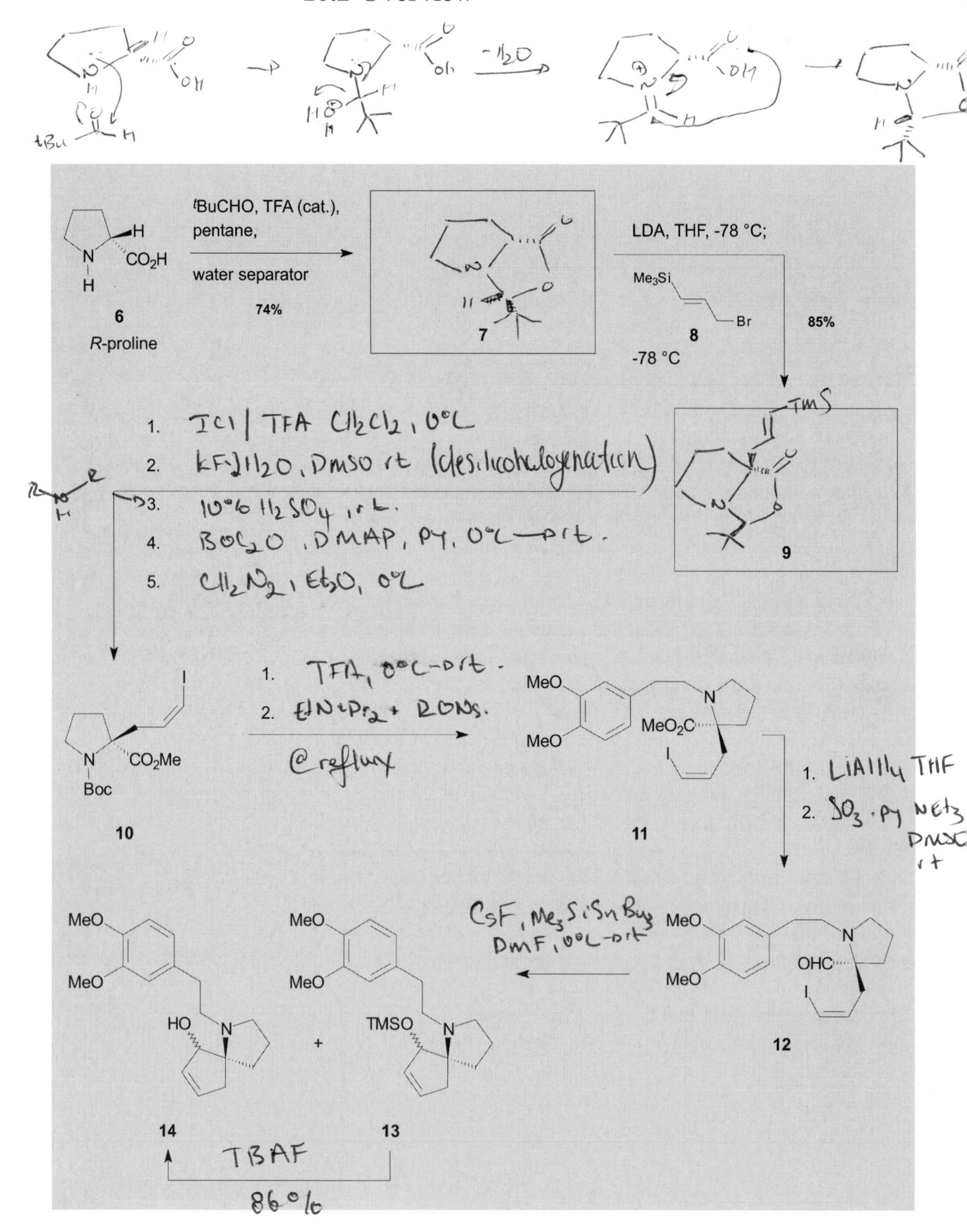

Friedle-Crafts alkylation of an allylic cation.

Poly phosphoric acid

MeO MeO HO N

14

PPA, 60°C

MeO MeO N H

15

1. BBr_3
2. CH_2Br_2, NaOH $PhCH_3$/H_2O, Aliquat 464 (PTC)

O O N H

16

0.05eq. OsO_4 Me_3NO, AcOH THF/H_2O.

O O N H HO OH

17

Swern variant DMSO, TFAA, CH_2Cl_2

O O N HO O

18

$HC(OMe)_3$, TsOH, CH_2Cl_2

O O N H O OMe

19

O O N H HO OMe

1

(-)-cephalotaxin

10.3 Synthesis

Problem

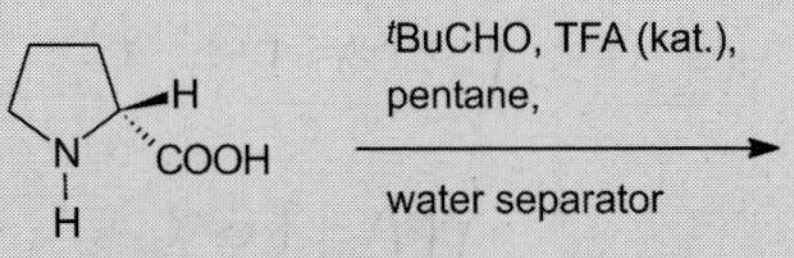

Tips

- Which is the most nucleophilic position in compound **6**, assuming that the amino acid is present in the neutral form depicted?
- The added aldehyde initially undergoes nucleophilic attack.
- Water is removed from the reaction mixture with the aid of a separator.
- Compound **7** is a bicyclic molecule.

Solution

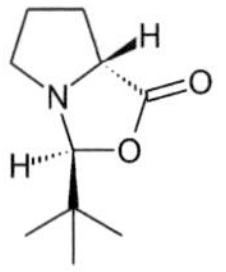

The transformation in question is a condensation reaction. A plausible mechanism suggests that the nitrogen atom from the neutral form of *R*-proline (**6**, present as one component in an equilibrium) attacks the carbonyl carbon atom of pivaldehyde. Elimination of water from the resulting hemiaminal leads to an iminium ion, which in turn reacts with the carboxyl group to give the *N*,*O*-acetal **7**.

A crucial factor in this synthetic strategy is the fact that only a single diastereomer is formed. The *tert*-butyl group in the bicyclic proline derivative is *cis* to the bridgehead hydrogen atom.

Discussion

Amino acids are often utilized as sources of chiral information in the synthesis of enantiomerically pure products. In the present synthesis, however, it is the abnormal enantiomer *R*-proline that is required, a compound 14 times as expensive as the natural *S* enantiomer.

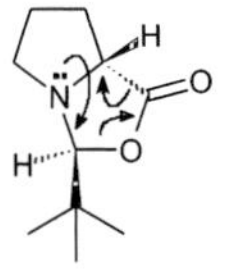

The use of pentane as a transport medium for water is unusual. Reaction actually occurs more rapidly in the presence of effective azeotroping agents, such as cyclohexane, but the corresponding higher reaction temperatures lead to increased formation of byproducts that are not only difficult to remove but also catalyze decomposition of the product. Compound **7** is extremely sensitive to hydrolysis. It becomes cloudy immediately upon exposure to air through the formation of insoluble proline. Carbon dioxide is released at temperatures above 100 °C in a 1,3-dipolar cycloreversion. For this reason considerable caution is required in the purification of **7**. Crystallization should be carried out from pentane at –40 °C. Distillation is possible at 90 °C and 0.07 mbar.[11]

Problem

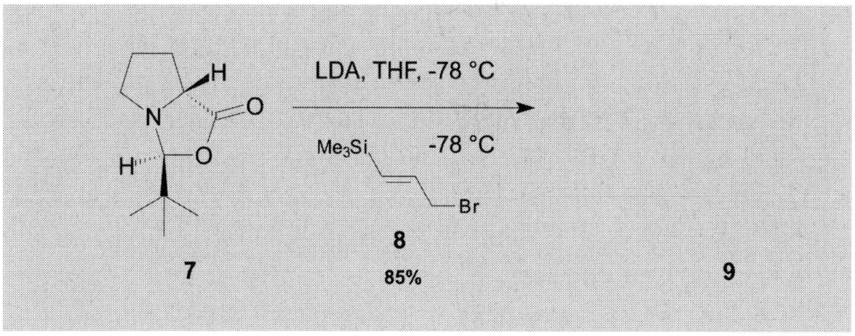
H
N
O
H
O
LDA, THF, -78 °C
Me3Si
-78 °C
Br
8
85%
7
9

Tips

- The most acidic position in compound **7** is deprotonated with LDA.
- Allylic bromide **8** functions as an alkylating agent.
- The *tert*-butyl group and the newly introduced substituent are *cis* to each other in product **9**.

Solution

In bicyclic system **7** the bridgehead hydrogen atom is made acidic by the neighboring carbonyl group. Deprotonation with LDA results in enolate **20**, the nucleophilic center of which is attacked from the *si* side. This *lk*-1,3-induction causes the *tert*-butyl group and the newly introduced allylic group to be *cis* to each other in product **9**.

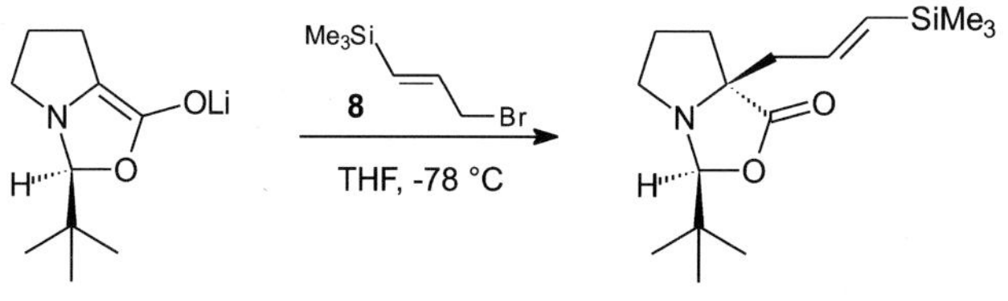
N
OLi
H
O
Me3Si
8
Br
THF, -78 °C
SiMe3
N
O
H
O

20 **9**

Problem

Propose a synthetic approach for the preparation of allylic bromide **8**.

Tips

- The starting material is propargyl alcohol.
- The reagent used to introduce the trimethylsilyl group is inexpensive. Protection of the alcohol function prior to silylation is unnecessary since the TMS group can in this case be removed selectively.
- The triple bond must be reduced to a double bond. Which of the following reducing systems is appropriate: a) H_2, *Raney* nickel; b) H_2, Pd/C; c) Red-Al®?
- Bromides can be prepared from alcohols in a one-step process.

Solution

1. 2.8 eq EtMgBr, THF, 10 °C
2. 2.8 eq TMSCl, 5 °C→ 70 °C
3. 1.1 eq 1.4M H_2SO_4, 45 °C

OH **21** → (94%) Me$_3$Si, OH **22**

$NaAlH_2(OCH_2CH_2OCH_3)_2$, Et_2O/toluene, 20 (71%)

Me$_3$Si, OH **23**

38% PBr_3, Py, Et_2O, reflux

Me$_3$Si, Br **8**

The dianion of propargyl alcohol **21** is disilylated with trimethylsilyl chloride (a process that occurs in somewhat higher yield with use of the magnesium rather than the lithium salt) and the silyl ether is subsequently hydrolyzed. The reagent for *trans* reduction of propargylic alcohol **22**, sodium bis(2-methoxyethoxy)aluminum hydride, is available under the trade name Red-Al® and is referred to by the abbreviations SDMA and SMEAH (for the mechanism of the reaction see Chapter 13). The other two reduction methods proposed above would result in *cis* hydrogenation of the triple bond.[12] Transformation of alcohol **23** into bromide **8** is accomplished with phosphorus tribromide[13] or with triphenylphosphine and tetrabromomethane.[14]

Problem

9 → 1. 2. 3. 4. 5. → 10

Tips

- Functional group manipulations on the vinylsilane are first conducted, followed by transformations affecting the bicyclic system.
- Iodine is introduced via a two-step addition-elimination mechanism.
- The reagent in the first step is iodine monochloride, which adds to the double bond.
- Fluoride ion accomplishes a desilicohalogenation in the second step.
- Intermediate **24** retains the *N*,*O*-acetal. The third step is the reverse of formation of this linkage.
- After acidic acetal cleavage the α-alkylated amino acid is protected with *N*-Boc in a fourth step and esterified in a fifth.
- Introduction of the Boc protecting group is usually accomplished with the anhydride of the corresponding acid.
- The reagent for preparation of the methyl ester is added in ether solution. In pure form it is a yellow gas (bp –23 °C) and highly explosive.

24

Solution

Iodine monochloride is first added to the double bond (in the sense of I^+ Cl^-) to give intermediate **25**. The β-effect of the silicon atom (i.e., stabilization of a positive charge on an atom in the β position) determines the regiochemistry of attack. For the subsequent desilicohalogenation reaction, fluoride proves to be the most suitable nucleophile. The adjacent scheme illustrates why an *E*-vinylsilane is transformed in this way into a *Z*-vinylic iodide. It has been established experimentally that the *E*/*Z* selectivity with *E*-vinylsilanes is greater than with the analogous *Z*-configured systems. Moreover, primary residues R, as in the present case, are better suited than secondary or especially tertiary analogs.[15]

9 → + ICl → 25a ⇌ ($F^⊖$) 25b → 24

Whereas compound **7**, as previously noted, is extremely sensitive to hydrolysis, it is actually rather difficult to cleave the *N*,*O*-acetal in the α-alkylated product **24**. Use of aqueous hydrogen bromide as suggested by *Seebach*[16] leads to only traces of product, although the desired result is achievable with dilute sulfuric acid.

The most common reagent for introducing the *tert*-butoxycarbonyl (Boc) protecting group is the pyrocarbonate Boc_2O. This is employed here in aqueous alkaline medium to which has been added the solubilizing agent dioxane.

The tip regarding the physical state of the reagent for esterification rules out the important methylating agents methyl iodide and dimethyl sulfate. Diazomethane is the mildest reagent available, and in ether solution it is relatively stable and easy to employ. It is prepared from *N*-nitroso compounds, the currently preferred starting material being *N*-methyl-*N*-nitroso-*p*-toluene sulfonamide. An ethereal solution of the sulfonamide is treated with alcoholic potassium hydroxide at 60 °C and the resulting diazomethane is distilled off together with diethyl ether.[17]

1. ICl, TFA, CH_2Cl_2, 0 °C.
2. $KF \cdot 2H_2O$, DMSO, RT, 84% over two steps.
3. 10% H_2SO_4, RT.
4. Boc_2O, NaOH, H_2O/dioxane, 0 °C → RT.
5. CH_2N_2, Et_2O, 0 °C, 97% over three steps.

Discussion

Overall, the consideration to this point has been devoted to a method for α-alkylation of amino acids. The problem derives from the fact that the stereogenic center in an amino acid is transformed into an sp^2 center of an enolate in the course of deprotonation. The trick involved in *Seebach*'s method is substrate-controlled creation of a second stereogenic center that remains intact during enolate formation and induces selectivity in the alkylation. The overall process is referred to as "self-regeneration of stereocenters" (SRS).[16]

Problem

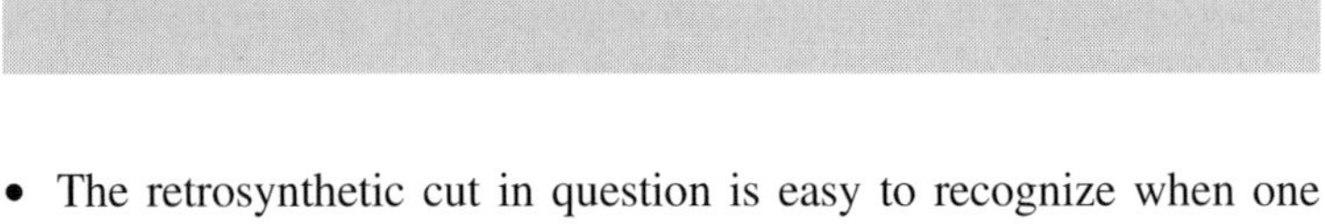

Tips

- The retrosynthetic cut in question is easy to recognize when one examines product **11** to find the fragment derived from starting material **10**.
- The second step is alkylation of a secondary amine.
- Which approach would be suitable for cleaving the Boc protecting group: a) NaH; b) TFA; c) H_2, *Raney* nickel; d) $NaBH_4$; or e) $KMnO_4$?

Solution

Cleavage of a *tert*-butoxycarbonyl group is usually accomplished with trifluoroacetic acid, either undiluted or (as here) in methylene chloride. By contrast, the Boc group is stable to oxidizing and reducing agents as well as base.[18] A convenient choice for the alkylation is the nosylate **26**.

1. TFA, CH_2Cl_2, 0 °C.
2. **26**, CH_3CN, NEt^iPr_2, reflux, 88% over two steps.

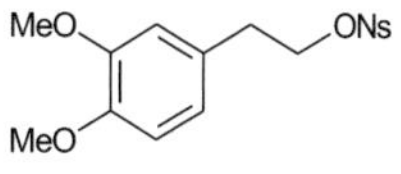

MeO MeO ONs **26**

Discussion

The nosylate group **30**, in comparison to a mesylate **27**, tosylate **28**, or benzene sulfonate **29**, displays a ten- to twenty-fold greater leaving tendency. If a leaving group of even greater reactivity is required, a triflate **31** or even a nonaflate **32** can be employed. Relative to mesylates (**27**) these are roughly 50000 to 150000 times more reactive.[19]

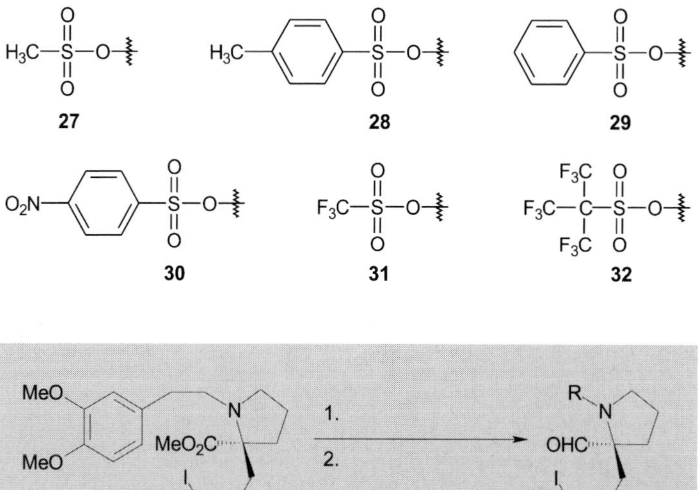

27 **28** **29** **30** **31** **32**

Problem

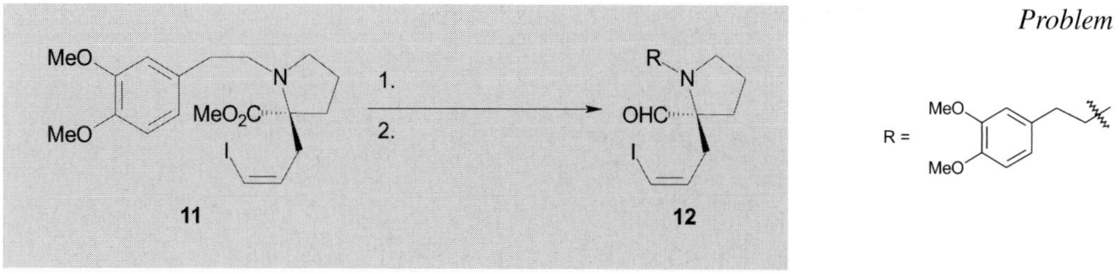

11 → 1. 2. → **12**

R =

Tips

- It is now possible to convert an ester into an aldehyde in a single step, but the safer route proceeds by way of reduction to an alcohol with subsequent oxidation.
- Esters can be reduced to alcohols with lithium aluminum hydride.

Solution

It is conceivable that the ester might be reduced with DIBAH or Red-Al® to an aldehyde in one step (see Chapter 3). Here the reduction was carried out with lithium aluminum hydride, however, producing an alcohol that was then oxidized to an aldehyde in a second step with activated dimethyl sulfoxide. Instead of the oxalyl chloride normally employed in a *Swern* oxidation, activation of the sulfoxide was accomplished in this case with the sulfur trioxide-pyridine complex (see Chapters 5 and 14).[20]

1. $LiAlH_4$, THF, –50 °C.
2. $SO_3 \cdot Py$, NEt_3, DMSO, RT, 81% over two steps.

Problem

R =

MeO

MeO

R N OHC I 12 1. TMSO R N 13 + HO R N 14 2.

Tips

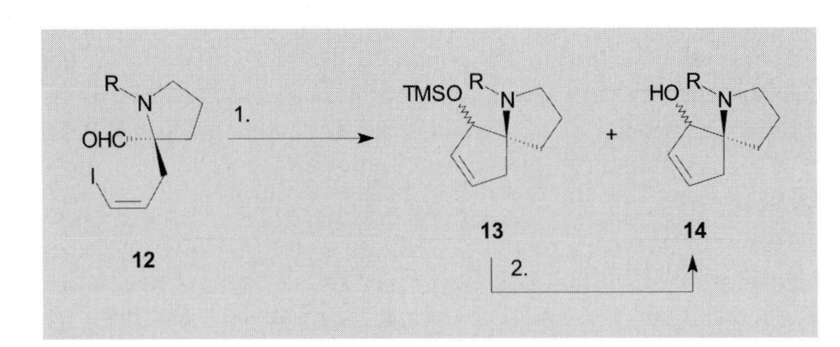

- The reagent for the first step is trimethylsilyl tributylstannane. Cesium fluoride is added as a source of fluoride.
- Byproduct **13** can be converted with TBAF into alcohol **14**.

Solution

Mori suggests the following mechanism for this key step in the synthesis:

$Bu_3Sn\text{-}SiMe_3$ **33** CsF $[Bu_3Sn\text{-}SiMe_3(F)]^{\ominus}$ $Cs^{\oplus}$ ≡ $X^{\ominus}Cs^{\oplus}$

34 $X^{\ominus}Cs^{\oplus}$ → **35** $[Bu_3Sn\text{-}SiMe_3(F)]^{\ominus}$ $Cs^{\oplus}$ → **36** $Bu_3Sn^{\ominus}$ $Cs^{\oplus}$ Si F → **37** $Cs^{\oplus}$ Bu_3SnI Si F → **38** OH

Fluoride ion first becomes coordinated with the silicon atom of tin reagent **33**. Further coordination with the carbonyl oxygen atom of substrate **34** leads to an especially activated hypervalent silicon species **35**,[21] from which a stannyl anion is eliminated. Halogen-metal exchange between the tin anion and vinylic iodide **36** produces vinylic anion **37**, which attacks in an intromolecular way the carbonyl group that has been activated through silicon coordination. Aqueous workup leads to alcohol **38**.

1. CsF, $Me_3SiSnBu_3$ (**33**), DMF, 0 °C → RT, 85% yield of **14** and 3% of **13**.
2. TBAF, 86%.

Discussion

This cyclization is a reaction developed by *Mori*, who carried out a series of investigations with tin reagent **33**. Fluoride is a much more reactive anion for initiation than any of the other halides, although it often leads to decomposition of the starting materials. In the case of vinylic or aryl halogen compounds the iodides are easier to transform than the bromides, and chlorides are unreactive. The carbonyl group can be derived from either an aldehyde or a ketone, and even esters are sufficiently electrophilic.[22]

Problem

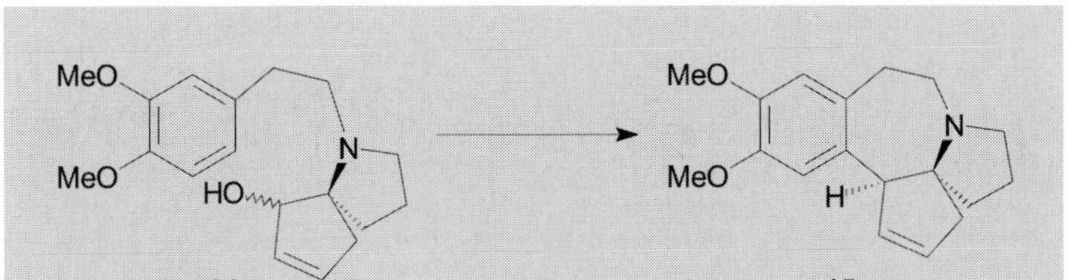

Tips

- This is a *Friedel-Crafts* alkylation with an allylic cation.
- The cation is obtained from an alcohol by the simplest route possible.

Solution

Heating to 60 °C in polyphosphoric acid leads to electrophilic aromatic alkylation.
PPA, 60 °C, 66%.

Problem

Tips

- At this point the two methoxy substituents must be transformed into the methylenedioxy group found in the natural product. Which reagents cleave aromatic methyl ethers: a) BBr_3; b) $BBr_3 \cdot SMe_2$; c) BCl_3; d) Me_3SiI; e) Me_3SiCl, NaI; f) EtSNa; g) NaCN?
- The methylenedioxy group is obtained from a 1,2-dihydroxyaromatic compounds in a double nucleophilic substitution reaction.

Solution

If boron tribromide is used for the ether cleavage, adduct **41** (from a combination of ether **39** and the electrophilic boron reagent) is attacked by bromide ion. The resulting intermediate **42** can than react further to phenol **43**.

In the second reaction, dibromomethane acts as the reagent. Since this requires working in a two-phase system, the phase-transfer catalyst Adogen 464 [methyltrialkyl(C_8-C_{10})ammonium chloride] is added.

1. BBr_3, CH_2Cl_2, –78 °C → RT; MeOH, 0 °C.
2. CH_2Br_2, NaOH, toluene/H_2O, Adogen 464, 0 °C → 100 °C, 51% over two steps.

Discussion

In principle there are two possibilities for cleaving a methyl aryl ether. Often one utilizes an electrophilic reagent that attacks the oxygen atom, thus initiating bond breakage. Especially valuable for this purpose are the reagents boron tribromide and boron trichloride, which cause ether cleavage under very mild conditions (temperatures as low as –78 °C). Trimethylsilyl iodide (less costly: TMSCl and sodium iodide) frequently shows sufficient reactivity only at higher temperature (65 °C).

The second possibility involves introduction of a strong nucleophile, which can engage in an S_N2 reaction at the methyl group with release of a phenolate anion. Reagents such as sodium thioethanolate in DMF or sodium cyanide in DMSO require reaction temperatures in excess of 100 °C and the presence of an electron-accepting substituent *para* to the methoxy group in the substrate in order to stabilize the resulting phenolate anion.

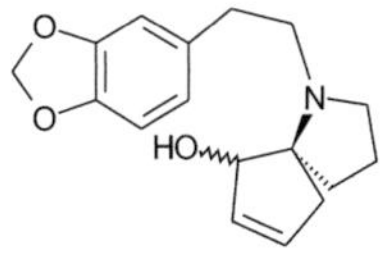

HO
44

The modest yield in the demethylation step must simply be accepted here because, surprisingly, *Friedel-Crafts* alkylation fails to occur with the methylenedioxy group present in **44**.[23] An explanation for this difference in reactivity has been provided by *Sha*.[24] Compared to the rotatability of the methoxy groups in **45**, the C-O bonds in methylenedioxy aromatic system **46** are rigidly fixed in a planar five-membered ring. This leads to reduced orbital overlap such that the –I effect (rather than a +M effect) of the oxygen atoms governs the reactivity of the benzene nucleus. In this way a tiny structural change has transformed an electron-rich aromatic system into one that is electron-poor.

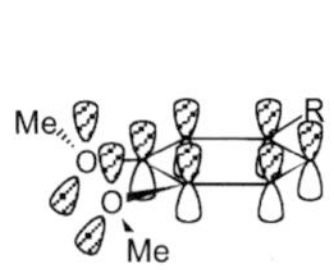

Me
Me
R
45

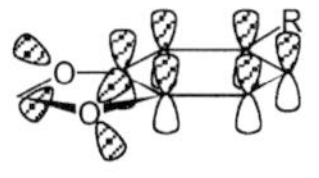

R
46

Problem

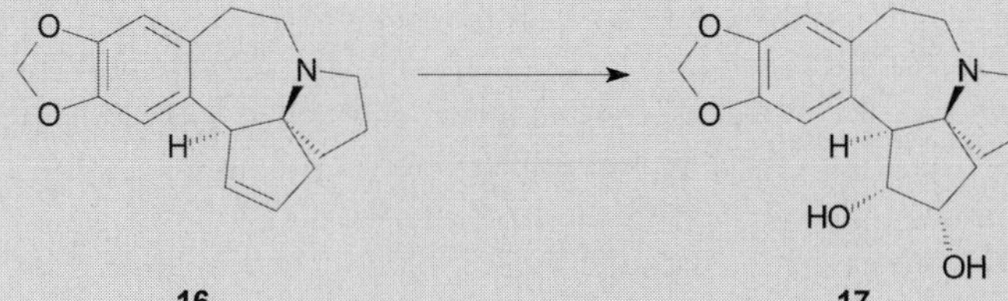

16 → 17

Tips

- The diastereoselectivity is in this case substrate-controlled.
- The reagent employed for dihydroxylation is expensive and toxic. It is therefore utilized in catalytic amounts together with a cooxidant.

Solution

The references to cost and toxicity point to osmium tetroxide as opposed to the inexpensive and convenient potassium permanganate.

The cooxidant in this case is trimethylamine-*N*-oxide, which often leads to higher yields with sterically hindered alkenes relative to the more usual *N*-methylmorpholine-*N*-oxide (NMO).[25]
0.05 eq. OsO_4, Me_3NO, AcOH, THF/H_2O, 0 °C → RT, 76%.

Problem

17 → **18**

Tips

- The enol and keto forms are in tautomeric equilibrium.
- Diols can be oxidized to diketones by the same methods used to convert primary alcohols into aldehydes.

Solution

The reaction here is a variant of the *Swern* oxidation (see Chapters 5 and 14).
DMSO, TFAA, CH_2Cl_2, –60 °C; NEt_3, –60 °C → 0 °C, 51%.

Discussion

There are several methods reported in the literature for transforming vicinal diols into α-diketones while avoiding the risk of C-C bond cleavage.[26] Examples include the standard *Swern* conditions (dimethyl sulfoxide and oxalyl chloride followed by triethylamine), or the use of DMSO activated by acetic anhydride, pyridine-sulfur trioxide complex, or dicyclohexylcarbodiimide (*Moffatt* oxidation). Diones are also obtained by treatment with benzalacetone as a hydride acceptor in the presence of catalytic amounts of tris(triphenylphosphine)ruthenium dichloride [$(PPh_3)_3RuCl_2$].[27] Recent developments include the use of *o*-iodoxybenzoic acid[28] or the oxoammonium salt of 4-acetamidotetramethylpiperidine-1-oxyl and *p*-toluenesulfonic acid.[29]

Problem

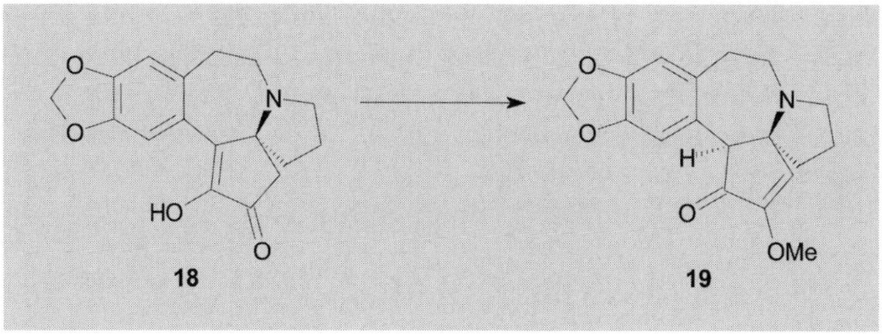
18 19

Tip

• At this point an enol ether is prepared.

Solution

$HC(OMe)_3$, TsOH, CH_2Cl_2, 47%.

Discussion

Compound **18** – in racemic form – was already known from the cephalotaxin synthesis of *Fuchs*,[30] so the steps beyond this needed only to be incorporated. However, when compound **18** in dioxane was heated under reflux with dimethoxypropane and *p*-toluenesulfonic acid, in analogy to the *Fuchs* work, methyl ether **19** was indeed obtained in 76% yield, but the product was found to be nearly racemic. Both acid-catalyzed and base-catalyzed racemization can be explained as follows:

Racemisierung, sauer katalysiert:

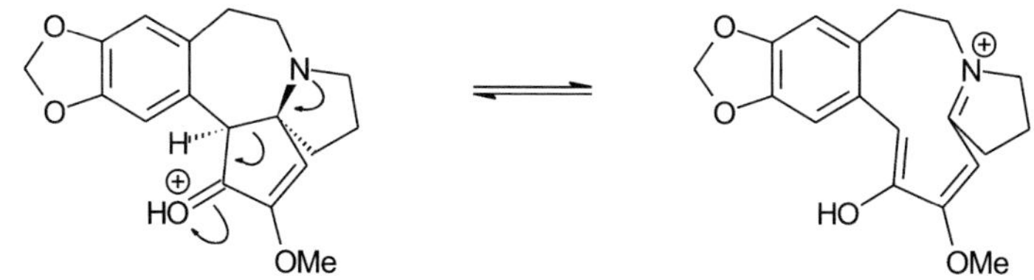
47 48

Racemisierung, basisch katalysiert:

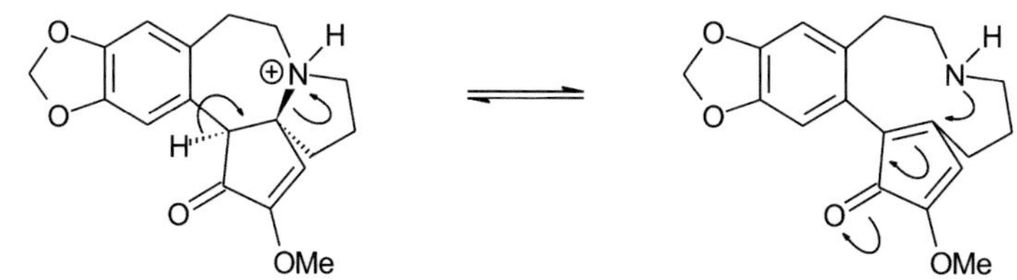
49 50

Other attempts to prepare the enol ether led to significantly lower yields: trimethylsilyldiazomethane and *Hünig* base produced 30% of compound **19**, whereas trimethylsilyl methyl ether or triethylsilyltrifluoromethane sulfonate gave only 7% or 13% of the product, respectively.

Problem

19

1

(-)-cephalotaxin

Tip

- The diastereoselective reduction is substrate-controlled.

Solution

The final step was also borrowed from the synthesis by *Fuchs*,[30] this time without unexpected complications.
$NaBH_4$, MeOH, –78 °C, 95%.

10.4 Summary

Starting with a chiral-pool compound, *Seebach*'s method for α-alkylation of amino acids was used to construct the decisive stereogenic center in the molecule. After coupling with the second building block **53** it proved possible to synthesize 1-azaspiro[4.4]nonane **51** with the aid of the *Mori* variation for generating vinylic anions. This step was followed by a *Friedel-Crafts* alkylation to give a benzazepan. Despite potential racemization problems, functional-group manipulation permitted synthesis of target molecule **1** in enantiomerically pure form. Starting with *R*-proline, the overall yield was 1.8% over 19 steps.

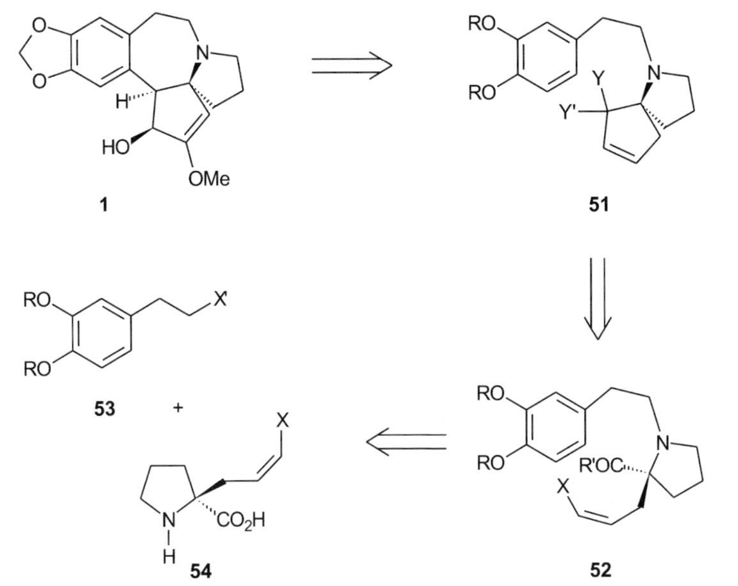

10.5 References

1 M.E. Wall, C.R. Eddy, T.T. Willaman, D.S. Correll, B.G. Schubert, H.S. Gentry, *J. Am. Pharm. Assoc.* **1954**, *43*, 503.

2 W.W. Paudler, G.I. Kerley, J. McKay, *J. Org. Chem.* **1963**, *28*, 2194.

3 D.J. Abraham, R.D. Rosenstein, E.L. McGandy, *Tetrahedron Lett.* **1969**, *10*, 4085.

4 S.K. Arora, R.B. Bates, R.A. Grady, R.G. Powell, *J. Org. Chem.* **1974**, *39*, 1269; S.K. Arora, R.B. Bates, R.A. Grady, G. Germain, J.P. Declercq, R.G. Powell, *J. Org. Chem.* **1976**, *41*, 551.

5 R.J. Parry, M.N.T. Chang, J.M. Schwab, B.M. Foxman, *J. Am. Chem. Soc.* **1980**, *102*, 1099.

6 R.G. Powell, D. Weisleder, C.R. Smith, W.K. Rohwedder, *Tetrahedron Lett.* **1970**, 815; R.G. Powell, D. Weisleder, C.R. Smith, *J. Pharm. Sci.* **1972**, *61*, 1227.

7 Cephalotaxus Research Coordinating Group, *Clin. Med. J.* **1976**, *2*, 263; Department of Pharmacology, Institute of Materia Medica, Chinese Academy of Medical Science, *Clin. Med. J.* **1977**, *3*, 131; J.Y. Zhou, D.L. Chen, Z. Shen, H.P. Koeffer, *Cancer Res.* **1990**, *50*, 2031.

8 J. Auerbach, S.M. Weinreb, *J. Am. Chem. Soc.* **1972**, *94*, 7172; M.F. Semmelhack, B.P. Chong, L.D. Jones, *J. Am. Chem. Soc.* **1972**, *94*, 8629; S.M. Weinreb, M.F. Semmelhack, *Acc. Chem. Res.* **1975**, *8*, 158.

9 L.F. Tietze, H. Schirok, *Angew. Chem. Int. Ed. Engl.* **1997**, *36,* 1124; *Angew. Chem. Int. Ed. Engl.* **1997**, *36*, 1124 und darin zitierte Literatur; T. Nagasaka, H. Sato, S. Saeki, *Tetrahedron Asymm.* **1997**, *8*, 191.

10 N. Isono, M. Mori, *J. Org. Chem.* **1995**, *60*, 115.

11 A.K. Beck, S. Blank, K. Job, D. Seebach, Th. Sommerfeld, *Org. Synth.* **1993**, *72*, 62.

12 T.K. Jones, S.E. Denmark, *Org. Synth.* **1985**, *64*, 182.

13 H. Hiemstra, W.J. Klaver, W.N. Speckamp, *Tetrahedron Lett.* **1986**, *27*, 1411.

14 L. Vidal, J. Royer, H.-P. Husson, *Tetrahedron Lett.* **1995**, *36*, 2991.

15 R.B. Miller, G. McGarvey, *Synth. Commun.* **1978**, *8*, 291.

16 D. Seebach, M. Boes, R. Naef, W.B. Schweizer, *J. Am. Chem. Soc.* **1983**, *105*, 5390; D. Seebach, A.R. Sting, M. Hoffmann, *Angew. Chem. Int. Ed. Engl.* **1996**, *35,* 2708; *Angew. Chem. Int. Ed. Engl.* **1996**, *35*, 2708.

17 T.J. De Boer, H.J. Backer, *Org. Synth. Coll. Vol. 4*, **1963**, 250; M. Hudlicky, *J. Org. Chem.* **1980**, *45*, 5377.

18 P.J. Kocienski, *Protecting Groups*, Georg Thieme Verlag, Stuttgart 1994, p. 192; T.W. Greene, P.G.M. Wuts, *Protective Groups in Organic Synthesis*, Wiley, New York 1991, p. 327.

19 P.J. Stang, M. Hanack, L.R. Subramanian, *Synthesis* **1982**, 85.

20 A.J. Mancuso, D. Swern, *Synthesis* **1981**, 165.

21 H. Sakurai, *Synlett* **1989**, *1*, 1.

22 M. Mori, N. Kaneta, N. Isono, M. Shibasaki, *Tetrahedron Lett.* **1991**, *32*, 6139; M. Mori, N. Kaneta, N. Isono, M. Shibasaki, *J. Organomet. Chem.* **1993**, *455*, 255; M. Mori, N. Kaneta, M. Shibasaki, *J. Organomet. Chem.* **1994**, *464*, 35; T. Honda, M. Mori, *Chem. Lett.* **1994**, 1013; A. Kinoshita, M. Mori, *Chem. Lett.* **1994**, 1475.

23 A similar cyclization with tin tetrachloride as Lewis acid and quantitative yields is given in: M.E. Kuehne, W.G. Bornmann, W.H. Parsons, T.D. Spitzer, J.F. Blount, J. Zubieta, *J. Org. Chem.* **1988**, *53*, 3439.

24 C.-K. Sha, J.-J. Young, C.-P. Yeh, S.-C. Chang, S.-L. Wang, *J. Org. Chem.* **1991**, *56*, 2694.

25 R. Ray, D.S. Matteson, *Tetrahedron Lett.* **1980**, *21*, 449.

26 C.M. Amon, M.G. Banwell, G.L. Gravatt, *J. Org. Chem.* **1987**, *52*, 4851.

27 S.L. Regen, G.M. Whitesides, *J. Org. Chem.* **1972**, *37*, 1832.

28 M. Frigerio, M. Santagostino, *Tetrahedron Lett.* **1994**, *35*, 8019.

29 M.G. Banwell, V.S. Bridges, J.R. Dupuche, S.L. Richards, J.M. Walter, *J. Org. Chem.* **1994**, *59*, 6339.

30 T.P. Burkholder, P.L. Fuchs, *J. Am. Chem. Soc.* **1988**, *110*, 2341.

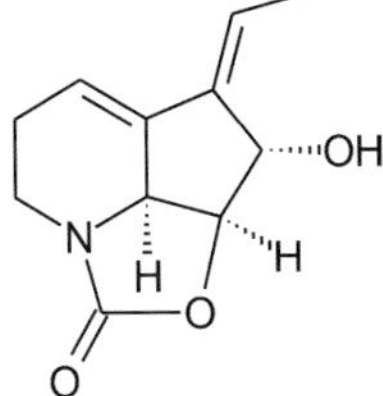

11

(+)-Streptazolin: Kibayashi (1996)

11.1 Introduction

Streptazolin (**1**), a secondary metabolite of *Streptomyces viridochromogenes*, was discovered and first isolated in 1981 by *Drautz* and *Zähner*[1] in the course of chemical screening. Isolation and purification of the antibiotic proved difficult, however, since in concentrated form it tends to polymerize.

The structure of streptazolin (**1**) was established both spectroscopically and through chemical degradation studies. That the absolute configuration is 2a*S*,3*S*,7b*S* was shown with the use of *Nakanishi*'s[2] chiral dibenzoate method on the degradation product **2** as well as by X-ray structural analysis[3] of *O*-acetyldihydrostreptazolin (**3**).

Streptazolin (**1**) is an antibiotic with a unique structure. It contains an unusual tricyclic ring system, a hexahydro-1*H*-1-pyrindine, along with an internal urethane unit and an exocyclic ethylidene sidechain.[4]

Streptazolin (**1**) shows limited antibacterial and fungicidal activity and influences the incorporation of thymidine in mouse spleen lymphocytes.

As a consequence of its interesting structure and promising pharmacological activity profile the substance offers an interesting challenge for organic synthesis.

The first racemic total synthesis was reported by *Kozikowsi*[5] in 1985, followed by the first enantioselective total synthesis by *Overman*[6] in 1987. In both of these syntheses the ethylidene sidechain was introduced with a *Wittig* reaction, which leads to a 1:2 mixture favoring the abnormal *E* isomer.

In 1996 *Kibayashi*[7] succeeded in completing an enantioselective total synthesis of the *Z* isomer of (+)-streptazolin (**1**) with a stereochemically defined exocyclic ethylidene sidechain.

11.2 Overview

HO, H, CO_2H, HO, CO_2H, H
4
L-tartaric acid
1. 2. 3. 4.
BnO, OBn, O, N, O, H
5
DEAD, PPh_3, THF, RT, 77%
Me_3Si, OH
6
7
1. $NaBH_4$, MeOH, 0 °C, 98%
2. Ac_2O, NEt_3, DMAP, CH_2Cl_2, RT, 95%
8
$BF_3\cdot OEt_2$, CH_2Cl_2, RT, 99%
$SiMe_3$, N, OBn, O, OBn
9
10
1. 2.
N, H, OBn, CHO, OBn, EtO_2C
11
CBr_4, PPh_3, CH_2Cl_2, RT, 98%
12

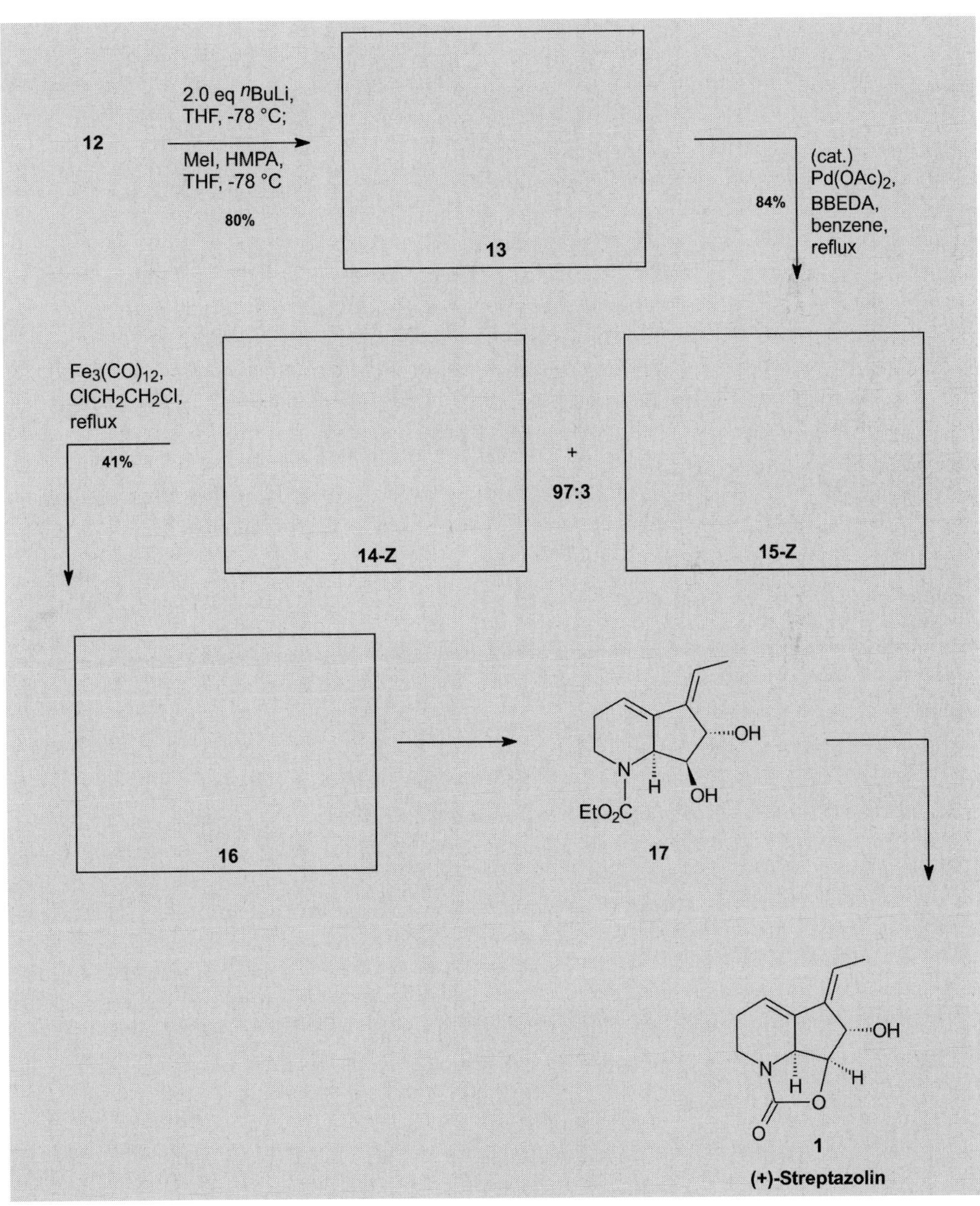
12
2.0 eq ^{n}BuLi, THF, -78 °C;
MeI, HMPA, THF, -78 °C
80%
13
84%
(cat.) Pd(OAc)$_2$, BBEDA, benzene, reflux
$Fe_3(CO)_{12}$, $ClCH_2CH_2Cl$, reflux
41%
14-Z
+
97:3
15-Z
16
OH
N
H
OH
EtO_2C
17
OH
N
H
H
O
O
1
(+)-Streptazolin

11.3 Synthesis

Problem

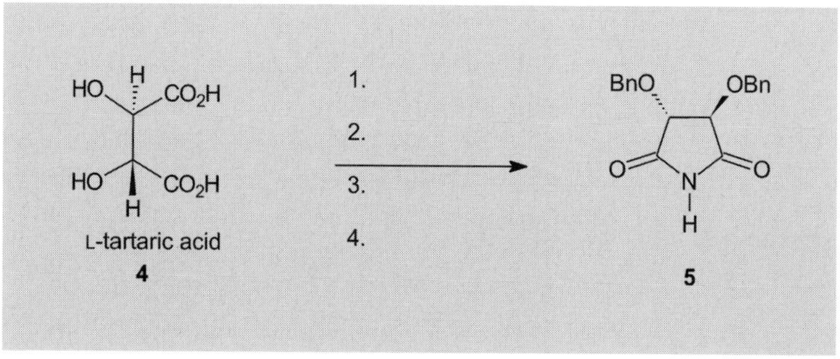

HO CO$_2$H / HO CO$_2$H
L-tartaric acid **4**
1. 2. 3. 4.
5

Tips

- This particular synthetic sequence includes not only a ring closure but also a series of protecting-group operations.
- The first step is an esterification with ethanol.
- In the second step the two hydroxy groups are protected.
- The third step is a ring closure. This involves prior saponification of the ester function.
- The third step creates an anhydride.
- In the fourth step the cyclic anhydride is first reopened and the compound is then recyclized to the succinimide **5**.

Solution

In order to avoid side reactions during protection of the hydroxy groups, L-tartaric acid (**4**) is first converted into the corresponding diethyl tartrate.[8] This involves heating in chloroform containing ethanol in the presence of a highly acidic ion-exchange resin and using a water separator.

In the second step the hydroxy groups are benzylated in a procedure developed by *Yamamoto*.[9] In this approach the tartrate is first deprotonated with sodium hydride and then treated with tetrabutylammonium iodide, benzyl bromide, and a catalytic amount of 18-crown-6.

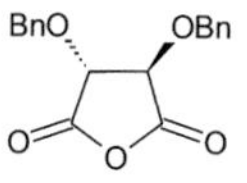

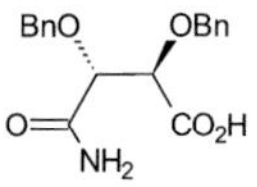

Cyclic anhydride **18** is created in the third step.[10] Thus, the ethyl ester is saponified with base and subsequently cyclized by way of the mixed anhydride from tartaric acid and acetyl chloride.

In the fourth step anhydride **18** is opened to succinamidic acid **19** by the introduction of ammonia, and this is again cyclized to give succinimide **5** via the mixed anhydride from **19** and acetyl chloride.

18

19

1. EtOH, Lewatit S 100, $CHCl_3$, reflux, 95%.
2. NaH, TBAI, BzBr, 18-crown-6, THF, $0\,^{\circ}C \rightarrow RT$, 76%.
3. 1 M NaOH, EtOH, $0\,^{\circ}C \rightarrow RT$; AcCl, reflux, 62%.
4. NH_3 (g), Et_2O, RT; AcCl, reflux, 91%.

Problem

BnO OBn
Me$_3$Si OH
6
DEAD, PPh_3,
THF, RT
77%
5 7

Tips

- Alcohol **6** is activated in preparation for a nucleophilic attack.
- Diethylazodicarboxylate (DEAD) and triphenylphosphine constitute a redox system.
- Overall, DEAD is reduced and triphenylphosphine is oxidized.
- Succinimide **5** is alkylated on the nitrogen atom.

Solution

Succinimide **5** is alkylated through a *Mitsunobu* alkylation[11] with (*Z*)-4-trimethylsilyl-3-butenol[12] (**6**), which leads to the *N*-butenylimide **7**.

Triphenylphosphine first attacks at the azo nitrogen atom of DEAD to produce the quaternary phosphonium salt **20**. This deprotonates succinimide **5** with the formation of intermediate **21**. Alcohol **6** reacts with **21**, in the course of which diethylhydrazine dicarboxylate **22** is formed along with the alkylated phosphine oxide **23**.

SiMe$_3$ OBn OBn 7

EtO—C(=O)—N=N—C(=O)—OEt + PPh_3 ⟶ EtO—C(=O)—N($\oplus PPh_3$)—N=C($O^{\ominus}$)—OEt

DEAD 20

Me$_3$Si OH 6 EtO—C(=O)—N(H)—N($\oplus PPh_3$)—C(=O)—OEt ⊖ 21 BnO OBn

BnO OBn 5

EtO—C(=O)—N(H)—N(H)—C(=O)—OEt + Me$_3$Si ... O—$\overset{\oplus}{P}Ph_3$ ⊖ OBn OBn

22 23

Compound **23** is the active alkylating agent. The α-carbon atom is activated toward nucleophilic attack, permitting it to alkylate the nitrogen of the deprotonated succinimide **5** to give *N*-butenylimide **7**. This releases triphenylphosphine oxide ($Ph_3P{=}O$). Thus, in the overall process DEAD is reduced to hydrazine derivative **22**, whereas triphenylphosphine is oxidized to triphenylphosphine oxide.

Problem

SiMe3 O ,,,OBn N OBn O **7**

1. $NaBH_4$, MeOH, 0 °C, 98%
2. Ac_2O, NEt_3, DMAP, CH_2Cl_2, RT, 95%

8

Tips

- Sodium borohydride is a reducing agent.
- Only one of the two carbonyl groups is attacked.
- The newly formed functional group undergoes reaction in the second step.

Solution

SiMe3 AcO ,,,OBn N OBn **8** O

N-Butenylimide **7** displays C_2 symmetry. The carbonyl groups are therefore homotopic, and it is irrelevant which of the two is reduced to a hydroxy group. The reduction products from both carbonyl groups are identical. Once reduction is accomplished, the resulting hydroxy group is acetylated to give acetoxylactam **8**.

Discussion

With respect to understanding topographic relationships, one usually examines structural fragments within a molecule. Thus, to establish the topicity of a substituent one ascertains what symmetry operation is capable of interconverting the two topographically related substituents. Alternatively, one can imagine exchanging sequentially the substituents of interest for a third group not yet present in the molecule. The nature of the resulting isomers provides direct information regarding the topicity class in question.

Homotopic substituents are interconverted by rotational axes. Substitution of either substituent leads to the same compound, so it follows that homotopic substituents are not subject to distinction.

Enantiotopic substituents are interchangeable through alternating axes of symmetry. Substitution leads to enantiomers, and therefore enantiotopic substituents are distinguishable under chiral conditions.

Diastereotopic substituents cannot be interchanged by carrying out any symmetry operation. Substitution leads to diastereomers, so diastereotopic substituents are distinguishable by chemical and physical means.[13]

Problem

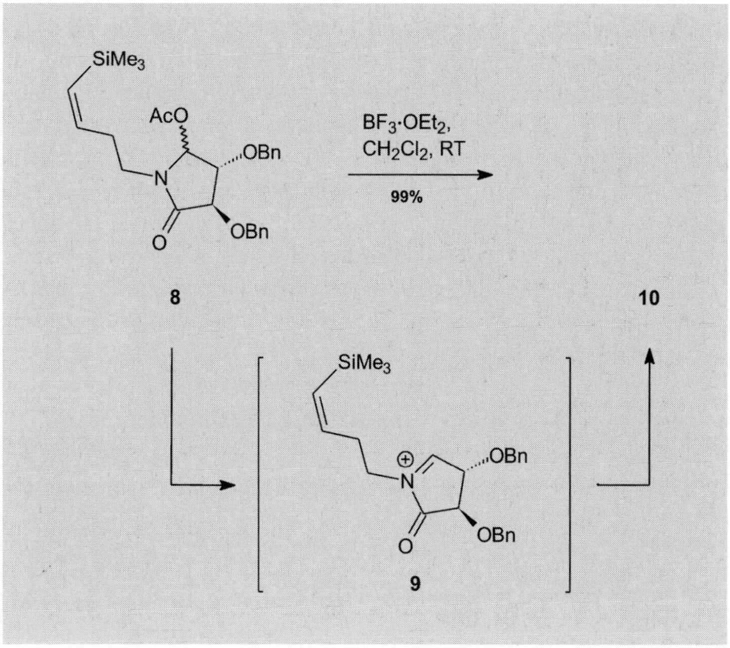

Tips

- The reaction takes place by way of the *N*-acyliminium ion **9**.
- This is a cyclization reaction.
- The product contains a bicyclic system.
- Cyclization begins with an iminium ion and concludes with a vinylsilane.
- The product forms stereoselectively.
- A six-membered ring results, specifically a 1,2,5,6-tetrahydropyridine.

Solution

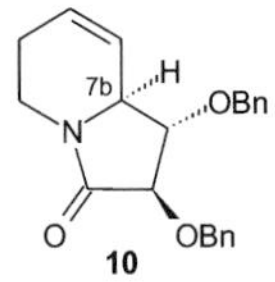

Upon addition of boron trifluoride etherate, acetoxylactam **8** eliminates an acetoxy group to produce *N*-acyliminium ion **9**. The indolizidinone **10** is formed diastereomerically pure in an iminium-ion-initiated cyclization reaction of the *Overman* type ending in a vinylsilane.[14]

Stereoinduction at C-7b occurs under substrate control. Because of steric shielding by the benzyloxy group in the α-position, the chain with the bulky (*Z*)-trimethylsilylvinyl residue attacks from above.

Overman[15] discusses two conceivable mechanisms for the cyclization. One possibility assumes a direct cyclization of iminium ion **9** via β-silyl cation intermediate **24** to the indolizidinone **10**. Cation **24** is stabilized by a β-effect of the silicon atom. Alternatively, iminium ion **9** might first undergo a charge-accelerated cationic aza-*Cope* rearrangement to allylsilaniminium ion **25**, which would then cyclize to **10** with loss of a silyl cation.

More probable is the second route, because *Overman* was able to demonstrate that the cationic aza-*Cope* rearrangement occurs more rapidly than cyclization. To this end, substrate **26** was treated with paraformaldehyde and camphorsulfonic acid, which led not to the tetrahydropyridine **27**, but rather to pyrrolidine **29**, formed by intramolecular *Mannich* cyclization of allylsilaniminium ion **28**.

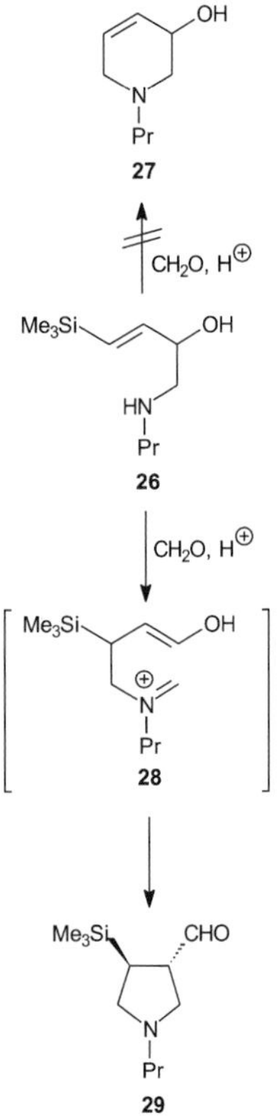

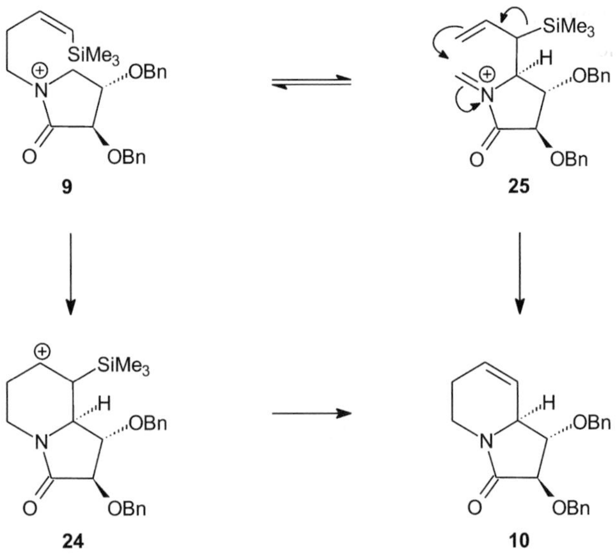

Discussion

1,2,5,6-tetrahydrophridine ring systems like that in **10** are found in various natural products and numerous pharmacologically active substances. Electrophilic cyclization reactions of iminium ions constitute an important method for constructing *N*-heterocyles. An advantage of the iminium-ion-initiated, vinylsilane-terminated cyclization is development of the ring system with complete regiocontrol of the location of the double bond. Moreover, it is possible to synthesize in this way 1-aryl-substituted tetrahydropyridines, which are not generally accessible from pyridine precursors.

Problem

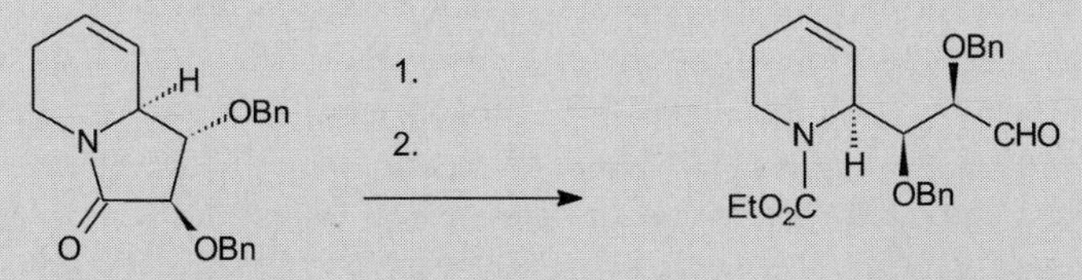

1.
2.

10 → **11**

Tips

- In this reaction sequence a ring is reductively cleaved and a protecting-group operation is accomplished.
- The reagent in question is capable of reducing tertiary amides selectively to aldehydes.
- A hemiaminal arises as an intermediate.

Solution

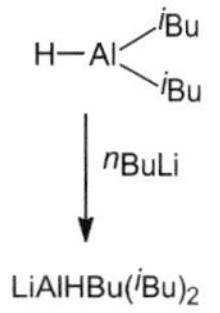

In the first step, the tertiary amide structure is reduced to hemiaminal **31** with the aid of the *at* complex (**30**) from DIBAH and butyllithium. The hemiaminal opens selectively to aminoaldehyde **32**. In a second step the ring nitrogen atom is protected as carbamate **11** with ethyl chloroformate.

1. DIBAH, ^{n}BuLi, THF, RT.
2. $ClCO_2Et$, CH_2Cl_2, RT, 86% starting from **10**.

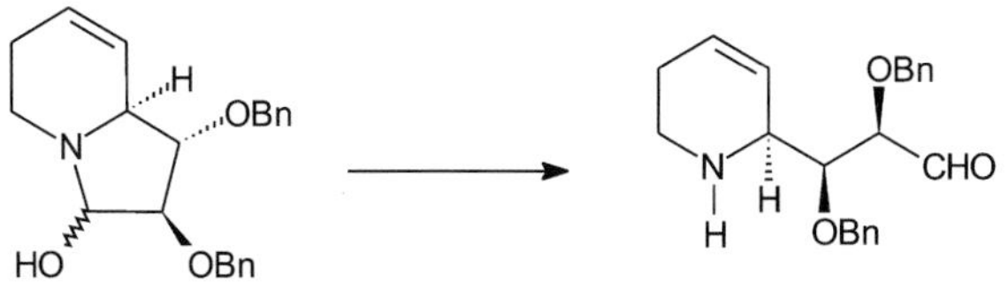

$H{-}Al(^{i}Bu)_2$ → (^{n}BuLi) → $LiAlHBu(^{i}Bu)_2$ **30**

31 → **32**

Discussion

The *at* complex from DIBAH and butyllithium is a selective reducing agent.[16] It is used for the 1,2-reduction of acyclic and cyclic enones. Esters and lactones are reduced at room temperature to alcohols, and at –78 °C to alcohols and aldehydes. Acid chlorides are rapidly reduced with excess reagent at –78 °C to alcohols, but a mixture of alcohols, aldehydes, and acid chlorides results from use of an equimolar amount of reagent at –78 °C. Acid anhydrides are reduced at –78 °C to alcohols and carboxylic acids. Carboxylic acids and both primary and secondary amides are inert at room temperature, whereas tertiary amides (as in the present case) are reduced between 0 °C and room temperature to aldehydes. The *at* complex rapidly reduces primary alkyl, benzylic, and allylic bromides, while tertiary alkyl and aryl halides are inert. Epoxides are reduced exclusively to the more highly substituted alcohols. Disulfides lead to thiols, but both sulfoxides and sulfones are inert. Moreover, this *at* complex from DIBAH and butyllithium is able to reduce ketones selectively in the presence of esters.

Problem

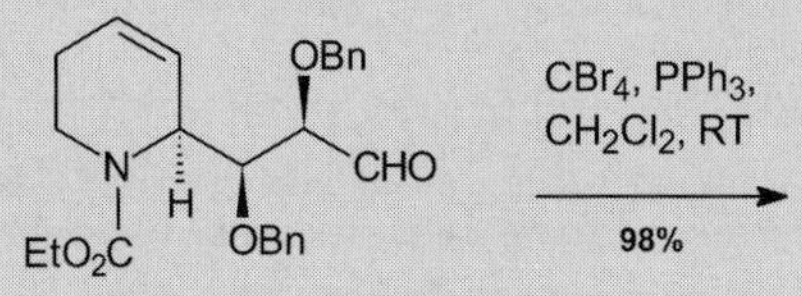

OBn, CHO, EtO_2C, N, H, OBn — **11** → (CBr_4, PPh_3, CH_2Cl_2, RT; 98%) → **12**

Tips

- Product **12** contains two bromine atoms.
- It is the aldehyde function that undergoes reaction.
- What analogous reaction also takes place with triphenylphosphine and an aldehyde group: a) the aldol reaction; b) the *Heck* reaction; or c) the *Wittig* reaction?
- Triphenylphosphine and carbon tetrabromide form an ylide.

Solution

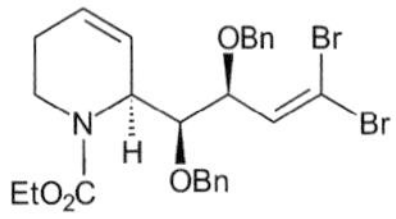

Analogous to the *Wittig* olefination, triphenylphosphine and carbon tetrabromide react to give an intermediate dibrominated phosphonium ylide. Ylide **33** extends the chain of aldehyde **11** with elimination of triphenylphosphine oxide to give 1,1-dibromoalkene **12**.[17]

OBn, Br, Br, N, H, EtO_2C, OBn — **12**

$$PPh_3 + CBr_4 \longrightarrow [Ph_3P{=}CBr_2] + Ph_3PBr_2$$

33

Problem

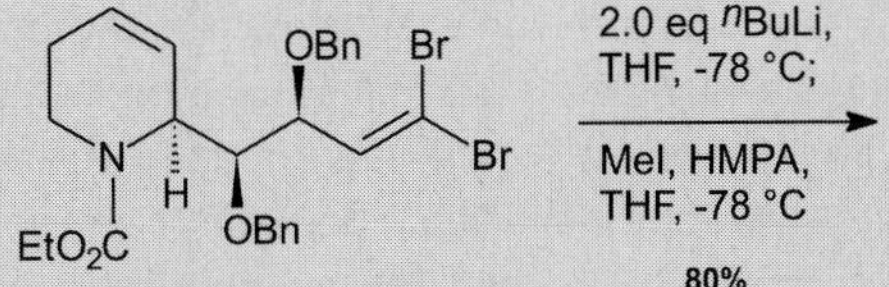

Tips

- The first step is a halogen-metal exchange reaction.
- A hydrogen atom shifts in a [1,2]-rearrangement.
- The product is an alkyne.

Solution

The reaction sequence **11** → **13** is known as the *Corey-Fuchs* process. This is generally understood to refer to the transformation of an aldehyde into an alkyne.[18]

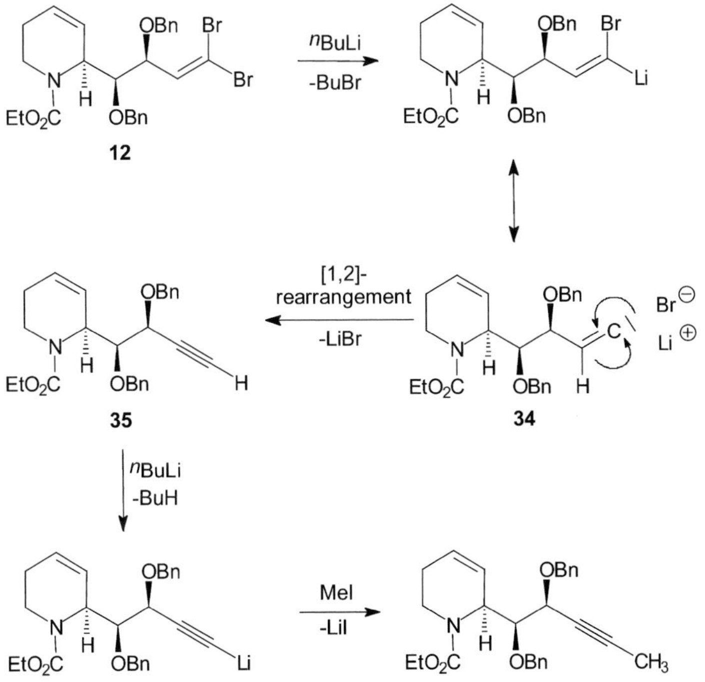

The first step is a halogen-metal exchange to give the α-lithiated bromoalkene **34**. Viewed from the perspective of its resonance structure, this can be regarded as a vinylic carbenoid species. Whether it rearranges as such or is first transformed into a free carbene is unclear. What follows is a [1,2]-rearrangement involving a hydrogen shift, leading to terminal alkyne **35**. This is so acidic that it reacts further with the second equivalent of butyllithium to give the corresponding lithium acetylide **36**. A final alkylation with methyl iodide leads to alkyne **13**.

Discussion

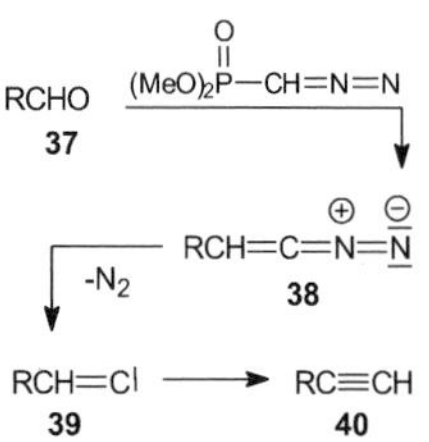

The one-step *Seyferth* procedure offers an alternative to the *Corey-Fuchs* sequence. After a *Horner-Wadsworth-Emmons* olefination of an aldehyde **37** to an unsaturated diazo compound **38**, elimination of nitrogen and [1,2]-rearrangement of the resulting vinylcarbene **39** leads here as well to a C_1-extended terminal alkyne **40**.[19]

Problem

BBEDA

13

84%

(cat.) Pd(OAc)$_2$, BBEDA, benzene, reflux

14-Z **97:3** **15-Z**

Tips

- This is a cyclization reaction.
- The product is bicyclic.
- A five-membered ring is created.
- The normal course of this palladium-catalyzed reaction is illustrated in the margin.

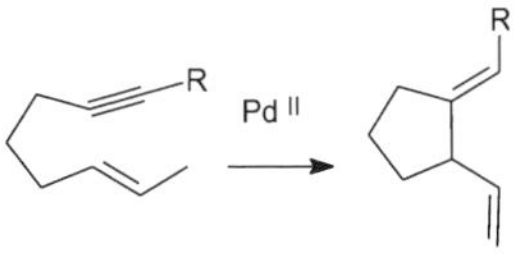

Pd II

Solution

A palladium-catalyzed 1,6-enyne cyclization of the *Trost*[20] type provides the desired bicyclic system as a mixture of hexahydro-5*H*-1-pyrindine **14-Z** and its double-bond isomer **15-Z** in a ratio of 97:3. The two can be separated by column chromatography. In the process, the exocyclic alkylidene sidechain is created with *Z* selectivity, and the rings are stereospecifically joined *cis*. It should be noted that only with the use of BBEDA [*N,N'*-bis(benzylidene)ethylene diamine] as a powerful σ-donor ligand is high *Z* selectivity achieved. Other catalytic systems, such as $Pd(OAc)_2$ or Pd(OAc) $_2(PPh_3)_2$/HOAc produce *E*/*Z* mixtures of **14** and **15** in ratios of 45:25:27:3 or 9:69:6:16 (**14-Z**: **15-Z**: **14-E**: **15-E**).

Trost discusses two possible mechanisms, in which complex **41** represents the active catalytic species. The BBEDA ligand serves to stabilize the hydridopalladium acetate formed from the substrates.[21]

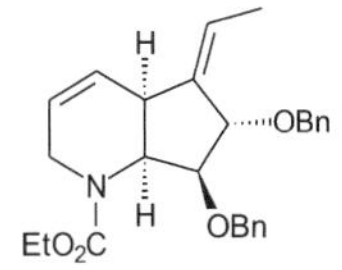

14-Z

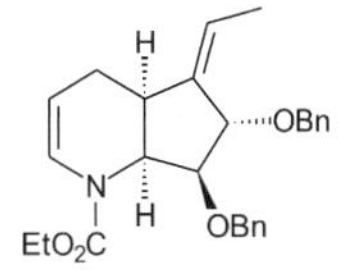

15-Z

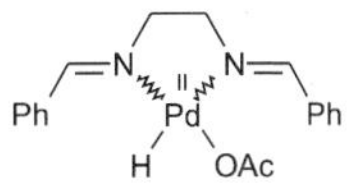

41

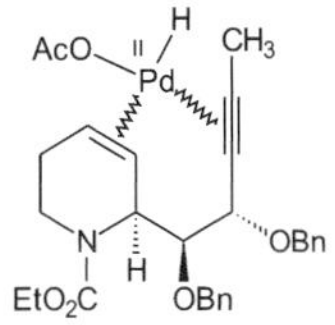

42

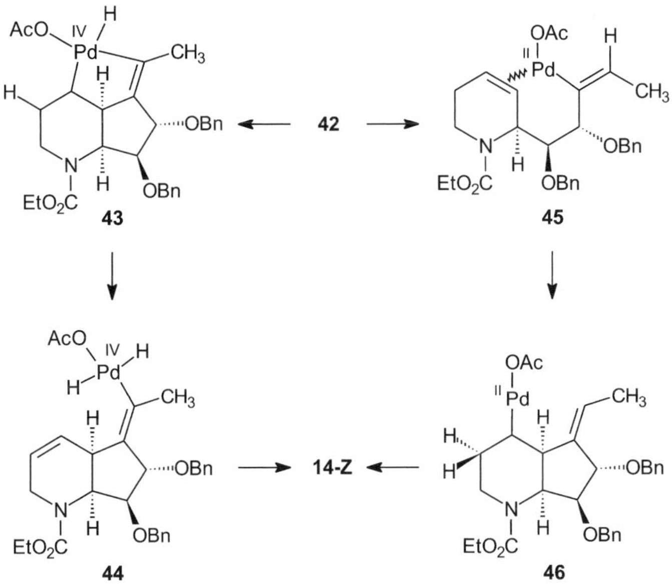

43 ← **42** → **45**

43 → **44** → **14-Z** ← **46** ← **45**

One conceivable pathway for the 1,6-enyne cyclization is a cyclopalladation via complex **42** to cyclic palladium compound **43**, which contains Pd^{IV}, followed by cleavage of the latter to intermediate **44** that subsequently releases hydridopalladium acetate and product **14-Z**. An alternative route proceeds by way of hydropalladation of the C-C triple bond to intermediate **45**, which after addition to the C-C double bond gives bicyclic system **46**. This leads to product **14-*Z*** following a β-hydride elimination. The appearance of double-bond isomeric compound **15**-*Z* is explained by another addition of the palladium species to the cyclic C-C double bond in **14-*Z*** and subsequent β-hydride elimination.

Problem

$Fe_3(CO)_{12}$, $ClCH_2CH_2Cl$, reflux

41%

14-Z → **16**

Tips

- A complex is formed, one in which triiron dodecacarbonyl is involved.
- This complex accomplishes an isomerization.

Solution

A stable tricarbonyl-(η^4-1,3-diene)-iron complex is formed.[22] This causes the 1,4-diene **14-Z** to isomerize to 1,3-diene **16**, which in turn leads to the hexahydro-1*H*-pyrindine present in streptazolin (**1**).

Problem

16 → **17**

Tips

- Cleavage of both benzyl protecting groups as well as the complex occurs in a single step.
- The reagent employed here is typically used in ether cleavages.
- It is a reagent that contains bromine atoms, and it is a strong Lewis acid.

Solution

Cleavage of the iron complex and the benzyl ether functions is accomplished with boron tribromide solution.
BBr_3, CH_2Cl_2, –90 °C, 51%.

Problem

17 → 1 (+)-streptazolin

Tips

- Cleavage of the carbamate and ring closure are also achieved in a single step.
- A lactone is created.

Solution

Ring closure of glycol **17** to the oxazolidinone ring in **1** takes place under basic conditions with sodium methoxide in methanol.
2.5% NaOMe, MeOH, reflux, 65%.

11.4 Summary

The key steps in the total synthesis of (*Z*)-(+)-streptazolin (**1**) are the stereoselective (C-7b) iminium ion-initiated cyclization of a vinylsilane using the *Overman*[14,15] procedure (**7** → **10**) together with a palladium-catalyzed *Trost*[20] enyne cyclization for selective construction of an exocyclic ethylidene side chain with the *Z* configuration (**13** → **14-Z**). The geometry of exocyclic alkenes is normally difficult to control. Whereas previous syntheses of streptazolin (**1**) employed *Wittig* reactions and led only to *E,Z* mixtures,[5,6] the enyne cyclization introduced by *Trost* offers the possibility of stereoselective reaction. Chiral information is derived from L-tartaric acid (**4**) and reflected in carbon atoms C-2a and C-3.

Considered retrosynthetically, streptazolin (**1**) is conveniently dissected by cleavage of the oxazolidinone ring into the bicyclic system hexahydro-5*H*-1-pyrindine **14-Z**, joined by a C-C bond. Ring opening in the bicyclic moiety provides the 1,6-enyne system **13**. Further disconnection with ring closure gives the C-N coupled bicyclic molecule **10**, an indolizidinone. This can be traced back to *N*-butenylimide **7** by opening of the tetrahydropyridine ring. Finally, imide **7** is derivable from L-tartaric acid (**4**).

3
7b
2a
OH
N
H
H
O
O
1

H
OBn
N
H
OBn
EtO_2C
14-Z

H
OBn
N
O
OBn
10

CH_3
N
H
OBn
EtO_2C
OBn
13

$SiMe_3$
O
OBn
N
O
OBn
7

HO
H
CO_2H
HO
CO_2H
H
4

Thus, (*Z*)-(+)-streptazolin (**1**) has been prepared stereoselectively in 17 steps with an overall yield of 2.2% starting from L-tartaric acid (**4**).

11.5 References

1 H. Drautz, H. Zähner, E. Kupfer, W. Keller-Schierlein, *Helv. Chim. Acta* **1981**, *64*, 1752; S. Grabley, P. Hammann, H. Kluge, J. Wink, P. Kricke, A. Zeeck, *J. Antibiot.* **1991**, *44*, 797; S. Grabley, P. Hammann, R. Thiericke, J. Wink, S. Philipps, A. Zeeck, *J. Antibiot.* **1993**, *46*, 343.

2 N. Harada, K. Nakanishi, *J. Am. Chem. Soc.* **1969**, *91*, 3989.

3 A. Karrer, M. Dobler, *Helv. Chim. Acta* **1982**, *65*, 1432. *Chemical Abtracts* Name: [2a*S*-(2aα,3α,4Z,7bα)]-4-ethyliden-2a,3,4,6,7,7b-hexahydro-3-hydroxy-1*H*-2-oxa-7a-azacyclopent-[*cd*]-inden-1-one.

4 Biosynthese: M. Mayer, R. Thiericke, *J. Org. Chem.* **1993**, *58*, 3486.

5 A.P. Kozikowski, P. Park, *J. Org. Chem.* **1984**, *49*, 1674; A.P. Kozikowski, P. Park, *J. Am. Chem. Soc.* **1985**, *107*, 1763; A.P. Kozikowski, P. Park, *J. Org. Chem.* **1990**, *55*, 4668.

6 C.J. Flann, L.E. Overman, *J. Am. Chem. Soc.* **1987**, *109*, 6115.

7 H. Yamada, S. Aoyagi, C. Kibayashi, *J. Am. Chem. Soc.* **1996**, *118*, 1054.

8 L.F. Tietze, T. Eicher, *Reaktionen und Synthesen*, Georg Thieme Verlag, Stuttgart 1991, p. 133.

9 H. Nemoto, S. Takamatsu, Y. Yamamoto, *J. Org. Chem.* **1991**, *56*, 1321.

10 J. Ohwada, Y. Inouye, M. Kimura, H. Kakisawa, *Bull. Chem. Soc. Jpn.* **1990**, *63*, 287.

11 O. Mitsunobu, *Synthesis* **1981**, 1.

12 C.J. Flann, T.C. Malone, L.E. Overman, *Org. Synth.* **1990**, *68*, 182.

13 B. Testa, *Principles of organic stereochemistry*, Verlag Chemie, Weinheim 1983.

14 T.A. Blumenkopf, L.E. Overman, *Chem. Rev.* **1986**, *86*, 857.

15 L.E. Overman, T.C. Malone, G.P. Meier, *J. Am. Chem. Soc.* **1983**, *105*, 6993; C.J. Flann, T.C. Malone, L.E. Overman, *J. Am. Chem. Soc.* **1987**, *109*, 6097.

16 S. Kim, K.H. Ahn, *J. Org. Chem.* **1984**, *49*, 1717.

17 F. Ramirez, N.B. Desai, N. McKelvie, *J. Am. Chem. Soc.* **1962**, *84*, 1745.

18 E.J. Corey, P.L. Fuchs, *Tetrahedron Lett.* **1972**, *36*, 3769.

19 R. Brückner, *Reaktionsmechanismen*, Spektrum Akademischer Verlag, Heidelberg 1996, p. 407.

20 B.M. Trost, *Acc. Chem. Res.* **1990**, *23*, 34; B.M. Trost, C. Pedregal, *J. Am. Chem. Soc.* **1992**, *114*, 7292; B.M. Trost, G.J. Tanoury, M. Lautens, C. Chan, D.T. MacPherson, *J. Am. Chem. Soc.* **1994**, *116*, 4255; B.M. Trost, D.L. Romero, F. Rise, *J. Am. Chem. Soc.* **1994**, *116*, 4268.

21 D.J. Rawlinson, G. Sosnovsky, *Synthesis* **1973**, 567; J.M. Davidson, C. Triggs, *J. Chem. Soc. A* **1968**, 1324.

22 W. McFarlane, G. Wilkinson, *Inorg. Synth.* **1966**, *8*, 181.

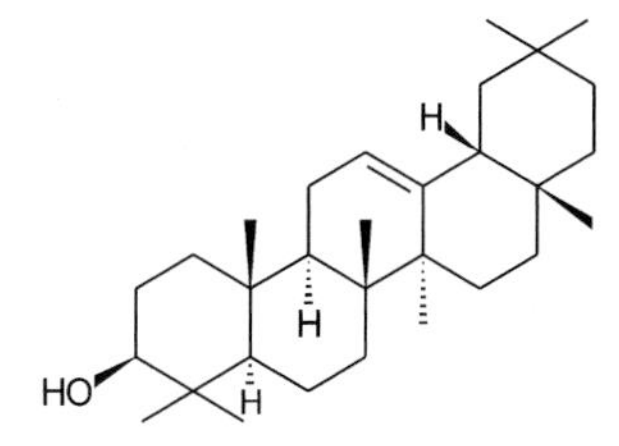

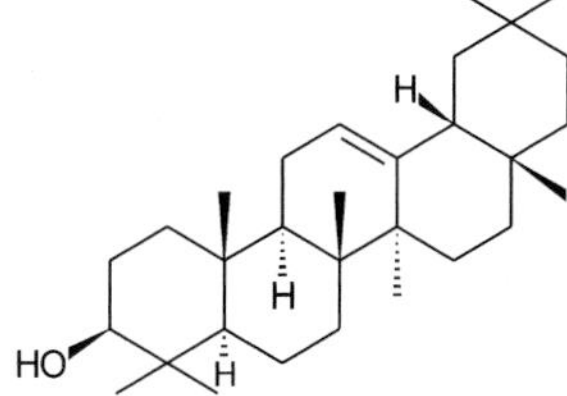

12

β-Amyrin: Corey (1993)

12.1 Introduction

β-Amyrin is a member of the pentacyclic triterpene family. It was first isolated from the sap of rubber trees of the species *Erythroxylum coca*.[1] Alongside erythrodiol (**2**), oleanic acid (**3**), and aegiceradienol (**4**), the title compound represents an interesting natural product structure that has been the target of numerous syntheses. As early as 1963 *Corey* reported the synthesis of an amyrin derivative.[2] The first totally synthetic access was reported by *Barton* in 1968 based on a sequence of 19 steps with an overall yield of ca. 0.001%.[3] Persistent interest in the cyclic triterpenes is reflected in the synthesis of δ-amyrin by *van Tamelen* in 1972[4] and of related pentacyclic substances by *Ireland* in 1976[5] and *Kametani* in 1978.[6]

1: R = CH_3
2: R = CH_2OH
3: R = COOH

4

Johnson in 1993 described an approach to racemic β-amyrin involving application of a biomimetic polyene cyclization.[7] In the same year *Corey* accomplished the enantioselective synthesis of compound **4**, a key intermediate that opened the way to stereoselective preparation of compounds **1**, **2**, and **3**.[8] A key step in the synthesis of β-amyrin (**1**) was the introduction of chiral oxazaborolidines for enantioselective carbonyl reduction. Based on these methods, generation of an enantiomerically pure epoxide and its stereoselective cationic cyclization led to the pentacyclic system of structure **1**. Diastereoselective cyclopropanation and an intramolecular protonation of a carbanion represent other interesting steps in this total synthesis.

12.2 Overview

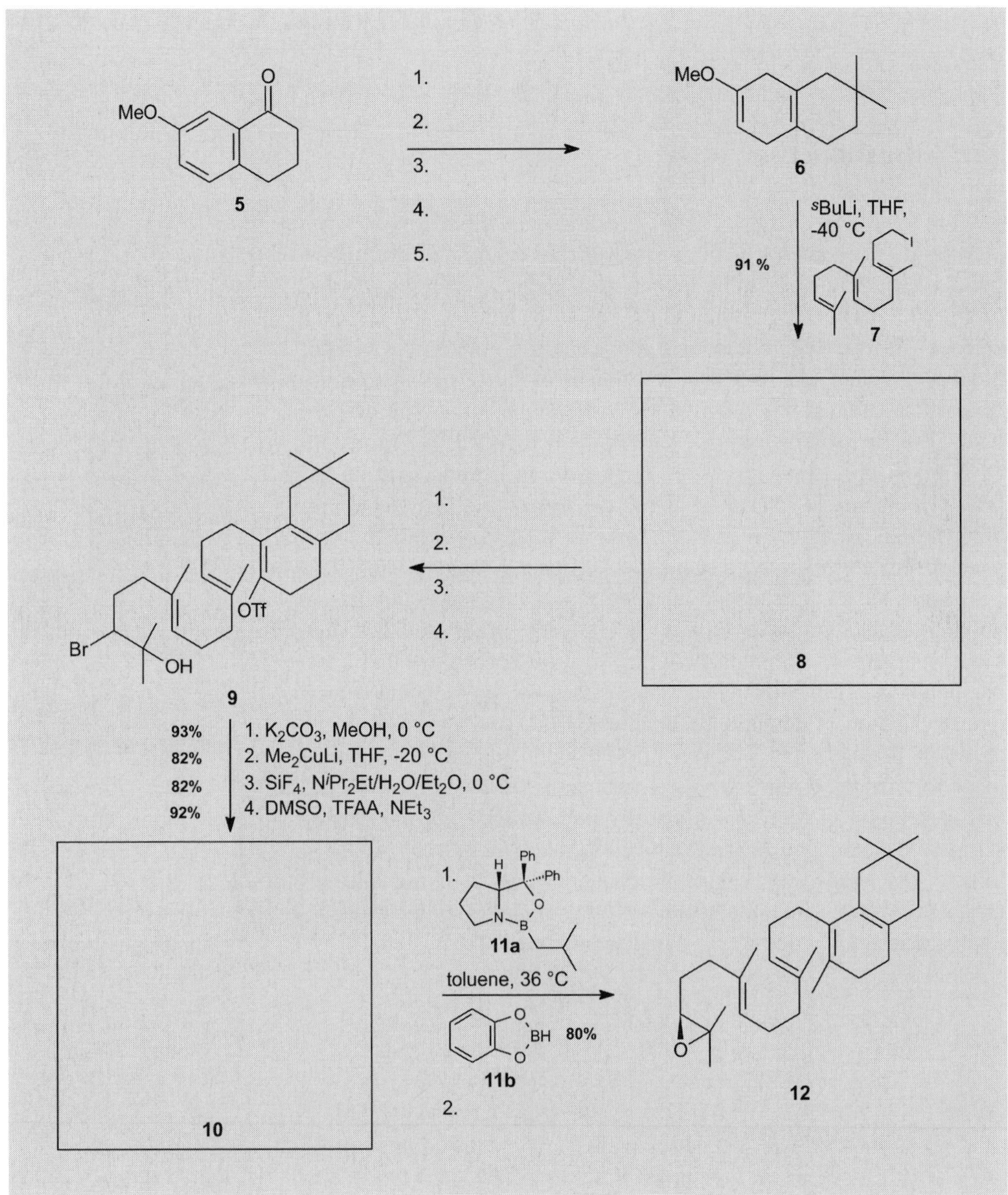

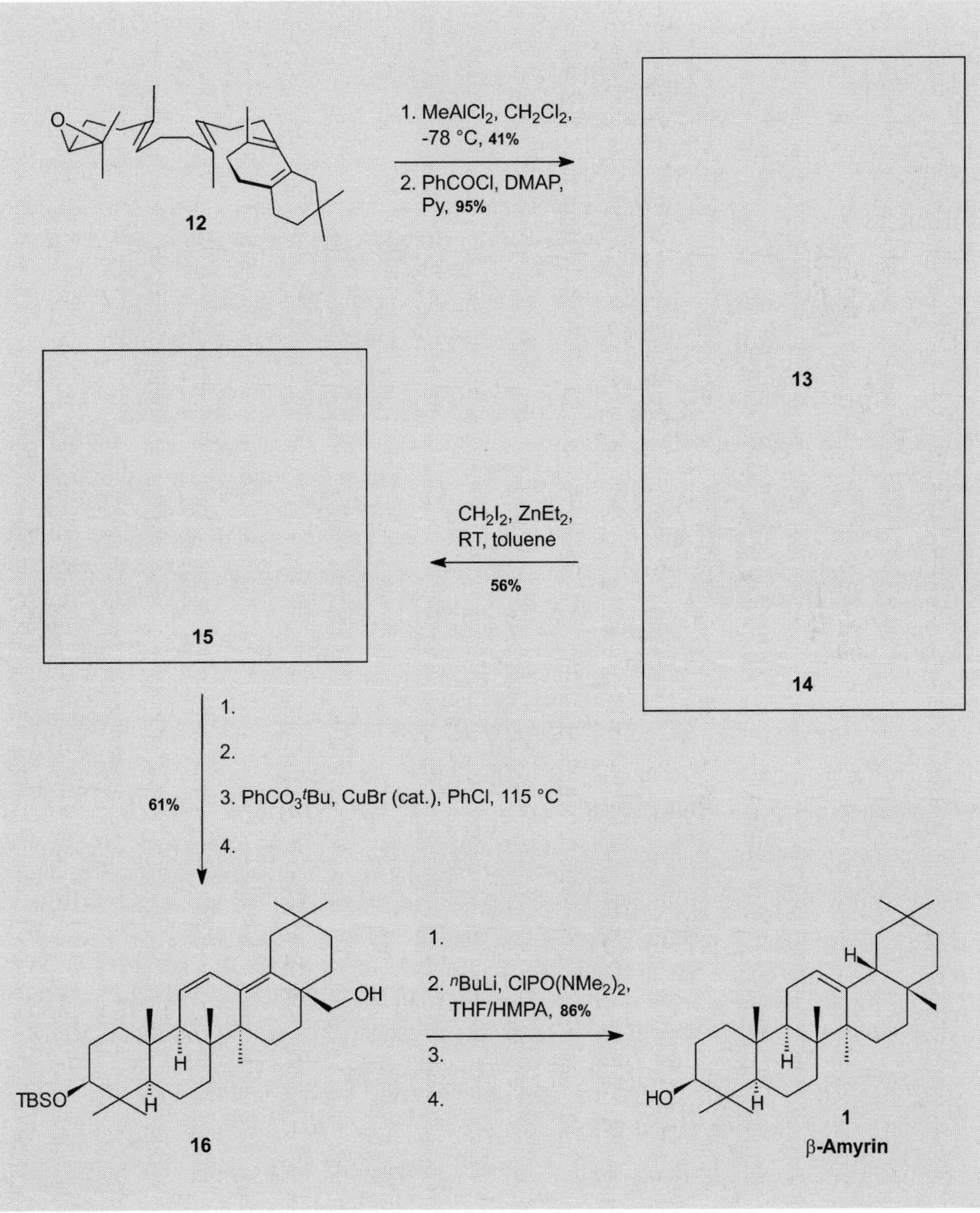

12
1. MeAlCl2, CH2Cl2, -78 °C, 41%
2. PhCOCl, DMAP, Py, 95%
13
CH2I2, ZnEt2, RT, toluene
56%
15
14
1.
2.
3. PhCO3tBu, CuBr (cat.), PhCl, 115 °C
4.
61%
OH
TBSO
H
16
1.
2. nBuLi, ClPO(NMe2)2, THF/HMPA, 86%
3.
4.
HO
1
β-Amyrin

12.3 Synthesis

Problem

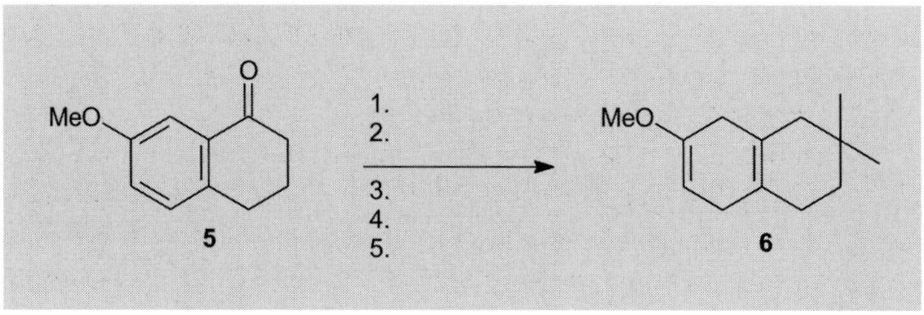

Tips

- The last transformation accomplishes the regioselective reduction of an aromatic ring. An alkali metal is utilized in this reaction.
- An ester group is displaced in the fourth step using a hydrogen nucleophile.
- Dialkylation and reduction occur in the first two steps.

Solution

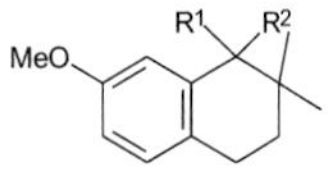

17: $R^1 = R^2 = O$
18: $R^1 = OH, R^2 = H$
19: $R^1 = OAc, R^2 = H$
20: $R^1 = H, R^2 = H$

At the outset, an α-dimethylation leads to compound **17**. Reduction of the ketone to secondary alcohol **18** and acetylation of the latter provides ester **19**. The ester group functions under acidic conditions as a leaving group, and it is replaced by a hydride anion with formation of compound **20**. The last step is a *Birch* reduction. These five steps were accomplished with an overall yield of 85%.

1. 2.2 eq. ^{t}BuOK, 4.0 eq. MeI, THF.
2. $NaBH_4$.
3. AcCl, Py.
3. TFA, Et_3SiH, CH_2Cl_2.
4. Li, ^{t}BuOH, NH_3(l)/THF, –40 °C.

Discussion

$$Li \xrightarrow{NH_3(l)} Li^{\oplus} + e^{\ominus}(NH_3)_n$$

Alkali and alkaline earth metals dissolve in liquid ammonia with the formation of solvated electrons. These solvated electrons constitute a very powerful reducing agent and permit reduction of numerous conjugated multiple-bond systems. The technique, named for *Birch*, provides selective access to 1,4-cyclohexadienes from substituted aromatics.[8] In the case of structures like **21** that are substituted with electron-donating groups, electron transfer produces a radical anion (here **22**) such that subsequent protonation occurs selectively in the *ortho* position (cf. intermediate **23**). A second electron-transfer step followed by another protonation leads to compound **24**.

donorsubstituierte Aromaten

OMe OMe OMe OMe

ROH, - RO⊖ ; e⊖; ROH, - RO⊖

21 22 23 24

akzeptorsubstituierte Aromaten

CO_2^- CO_2^- CO_2^- CO_2^-

ROH, - RO⊖ ; e⊖; ROH, - RO⊖

25 26 27 28

In an analogous way, electron transfer to an acceptor-substituted aromatic like **25** produces a radical anion of type **26**. This is protonated in the *ipso* position to give intermediate **27**. A second electron transfer and protonation leads similarly to product **28**. In most cases ^{t}BuOH proves to be a good source of protons.

Problem

MeO

6

7

^{s}BuLi, THF, -40 °C

91%

8

Tips

- This is a nucleophilic substitution reaction.
- Regioselective generation of the organolithium compound benefits from two stabilizing effects.

Solution

Deprotonation occurs very regioselectively at the allylic position and *ortho* to the methoxy group. Resonance stabilization of the anion by the double bond, together with coordinative stabilization of the lithium atom by the methoxy group, is responsible for selective formation of intermediate **29**.

Li

MeO

29

Discussion

Homofarnesyl iodide **7** was prepared by the reaction sequence shown in the margin. Of interest here is the two-step transformation of an alkyl halide into a C_1-extended alkyl halide.[9] Compound **30** is first subjected to a nucleophilic substitution by an organolithium species with formation of a homoallylic phenyl thioether. This is then methylated in a second step to an intermediate sulfonium salt. The final S_N2 reaction with an iodide ion releases thioanisol as a stable leaving group to give compound **7**.

Problem

Tips

- The first three reactions are used to construct an enol triflate.
- In addition to the desired ketone, acid hydrolysis also leads to another isomerization product.
- Use of 2-mercaptoethanol produces an intermediate that, with the aid of a noble metal, can be hydrolyzed to the desired ketone free of isomerization.
- The thermodynamically favored enol triflate is prepared in the third step at low temperature and in the presence of a strong base.
- What follows is an exceptionally chemoselective reaction for construction of the bromohydrin functionality.
- The bromine atom is derived from a brominating agent also used in radical brominations. The OH group comes from a supplemental solvent.

Solution

The enol ether is cleaved under acidic conditions in the first step to give ketone **32**. In order to suppress isomerization of the ketone to an α,β-unsaturated system, hydrolysis is conducted in the presence of 2-mercaptoethanol, which leads to formation of the S,O acetal **31**.

Silver ion makes possible in the second reaction an isomerization-free hydrolysis of the S,O acetal to ketone **32**. Generation of the enol triflate **33** is accomplished in the third step with the *Hendrickson-McMurry* reagent (Tf_2NPh).[10] Addition of an alcohol produces the potassium alkoxide, which because of its lower basicity permits isomerization to the thermodynamically favored enolate. Chemoselective reaction to bromohydrin **9** is achieved in the last step with NBS as brominating agent in aqueous THF. NBS acts here as a source of cationic bromine in an ionic mechanism. The intermediate bromonium ion forms preferentially at a) electron-rich double bonds and b) the sterically least hindered double bond. It also opens in such a way as to provide the most stable carbocation.

1. 5.0 eq. 2-thioethanol, 0.1 eq. TFA, CH_2Cl_2, 0 °C → 23 °C.
2. $AgNO_3$, Ag_2O, THF/CH_3CN/H_2O, 84%.
3. 1.15 eq. KHMDS, 0.15 eq. ^{t}AmOH, Tf_2NPh, THF, –78 °C → –35 °C, 86%.
4. NBS, THF/H_2O (5:1), 0 °C, 81%.

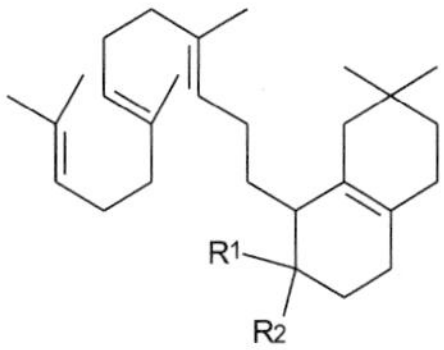

31: $R^1 = SCH_2CH_2O = R^2$
32: $R^1 = R^2 = O$

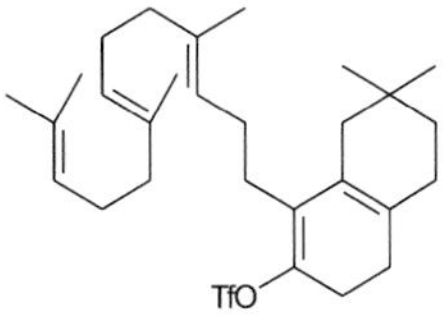

33

Problem

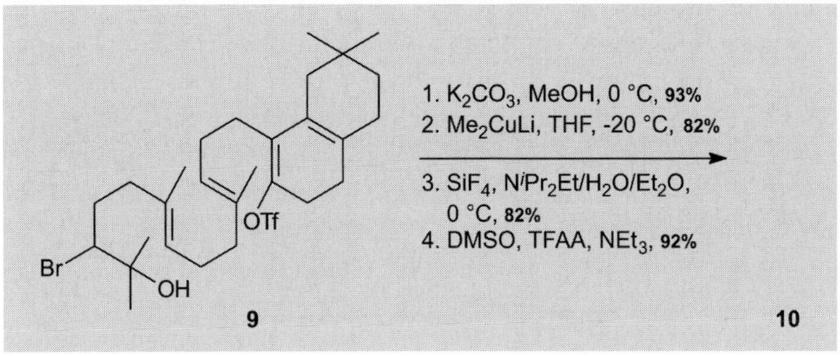

1. K_2CO_3, MeOH, 0 °C, 93%
2. Me_2CuLi, THF, -20 °C, 82%
3. SiF_4, $N^iPr_2Et/H_2O/Et_2O$, 0 °C, 82%
4. DMSO, TFAA, NEt_3, 92%

9 → **10**

Tips

- In the second reaction a methyl group is introduced with the aid of a cuprate. The bromine atom and the free alcohol interfere.
- Weakly basic conditions suffice to initiate an intramolecular S_N2 reaction in the first step.
- The last transformation accomplishes in a very mild way the oxidation of a functional group that has arisen from Lewis-acid opening of a group introduced in the first step.
- SiF_4 serves as a Lewis acid and permits simultaneous introduction of a fluorine atom.
- An epoxide is opened regioselectively to give the most stable carbocation.

Solution

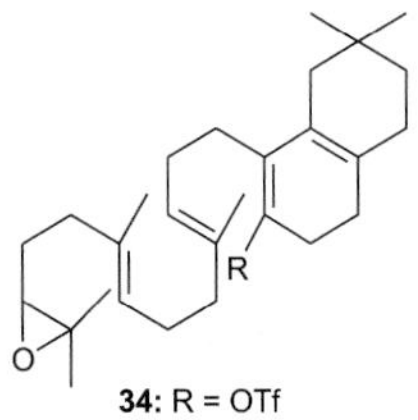

34: R = OTf
35: R = Me

The first reaction leads under basic conditions to cyclization to epoxide **34**. Replacement of the triflate group by a methyl group to give **35** is accomplished with the aid of a *Gilman* cuprate.[11] The epoxide function is not affected by this reaction. Tetrafluorosilane serves as both a Lewis acid and a fluoride source in the subsequent regioselective epoxide opening to compound **36**. SiF_4 reacts with water to give silicic acid and hydrogen fluoride, and the latter reacts with the *Hünig* base to produce an ammonium fluoride salt. Because of poor ion-pair formation by the ethyldiisopropylammonium cation, fluoride ions are present in the free state in solution and can thus act as nucleophiles. The final *Swern* oxidation with TFAA as activating agent for DMSO (see Chapter 1) leads to compound **10**.[13]

Discussion

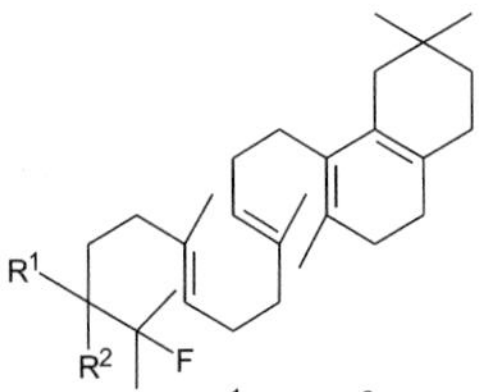

36: R^1 = OH, R^2 = H
10: R^1 = R^2 = O

Intramolecular S_N2 reaction of an enantiomerically pure halohydrin constitutes, alongside the related cyclization of an α-haloacetoxy compounds (see Chapter 13) and the asymmetric epoxidation introduced by *Jacobson*, a very elegant approach to the preparation of stereoisomerically pure epoxides.

Cuprates as sources of carbon nucleophiles are capable of displacing leaving groups at sp^2-hybridized carbon atoms. Epoxides can also be opened with organocopper compounds, but this only occurs at room temperature.[14] Selective reaction of dimethylcuprate with the triflate group is therefore possible by appropriate choice of the reaction temperature.

Problem

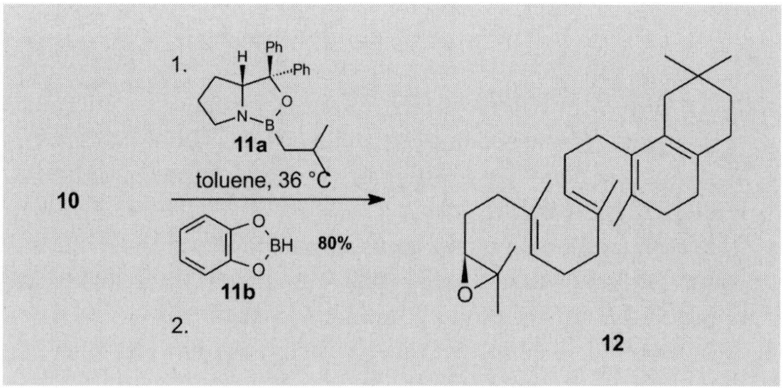

10 → 12
1. 11a, toluene, 36 °C, 11b, 80%
2.

Tips

- The first reaction is an asymmetric reduction with catecholborane (**11b**) as a hydride source.
- The boron-containing bicyclic compound is a chiral reagent containing both electrophilic and nucleophilic reaction centers.

- Coordination of the ketone takes place with a minimization of steric interactions at the boron atom. Compound **11b** coordinates as an electrophile with the nitrogen atom in compound **11a**.
- The product of the first transformation is an enantiomerically pure bromohydrin. The reaction conditions for the second reaction have been employed previously in the course of this synthesis.

Solution

The method *Corey* developed for catalytic asymmetric reduction of ketones depends on the use of a chiral oxazaborolidine such as compound **11a**.[15] These species, when incorporated into compounds with boron-containing reducing agents, act as chiral catalysts. The so-called CBS reduction[15] (*Corey-Bakshi-Shibata*) permits selective generation of chiral alcohol **37**. This then cyclizes stereoselectively under strongly basic conditions in an intramolecular S_N2 reaction to epoxide **12**.

1. 0.5 eq. **11a**, 2.5 eq. **11b**, toluene, 36 °C, 80%.
2. ^{i}PrOH, ^{i}PrONa, reflux, 83%, 92% *ee*.

Discussion

Oxazaborolidines (**39**), readily accessible from the amino acids (*R*)- or (*S*)-proline (**38**), have been described as “chemoenzymes” or “molecular robots” as a result of their mode of catalytic activity.

R-proline (**38**) → 1. MeOH, $H^{\oplus}$ 2. ArMgCl 3. R_3B; R = H, Me, ^{n}Bu, ^{i}Bu → **39**

The mechanism of enantioselective reduction here can be described in simplified form as follows:

Compound **40**, together with a hydride source (diborane, catecholborane), forms the catalytically active Lewis acid-base complex **41**. In the process, the electrophilic boron atom of the borane binds to the nucleophilic nitrogen atom of **41**. Strong coordination of the borane with the nitrogen atom of **40** and the resulting high degree of catalytic activity is explained by minimization of the ring strain in the 5/5 ring system. The latter is caused by a double-bond contribution to the B-N bond.

In the next step the carbonyl oxygen atom of the keto group to be reduced in **42** binds to the boron atom of **41**, accomplishing in structure **43** a very close alignment of the reaction centers. Prochiral carbonyl compound **42** arranges itself such that the steric interactions are minimized. That is to say, the smaller residue (R_s) points in the direction of the alkyl group on the boron atom in **43** and the larger residue (R_l) points in the opposite direction.

Subsequent transfer of a hydride ion occurs highly stereoselectively by way of a six-membered transition state to produce **44**. Bicyclic system **40** is regenerated from **44** with the elimination of compound **45**, and **40** reacts with the borane or with **44** to give the corresponding adduct **41** thereby completing the catalytic cycle. Chiral alcohol **46** is released by hydrolysis of **45**.

Ph Ph O N B BH3 R **41**

O Rk Rg **42**

R Ph Rk B O N O B H H H Rg Ph **43**

BH3 • THF or **44**

Ph Ph O N B **40** R

R Ph Rk B O O N H Rg BH2 Ph **44**

+

Rk OBH2 Rg H **45**

H⊕

OH Rk Rg **46**

B-Alkyl-substituted oxazaborolidines are preferred over *B*-H-substituted systems because of their stability and ease of preparation. The reaction itself usually occurs at room temperature in THF upon addition of the ketone to a solution of catalyst **40** (ca. 1–10 mol%) and the borane.

The most important criterion for achieving high enantioselectivity in this reaction is a sufficient steric difference between the residues R_s and R_l. It is also noteworthy that an increase in enantioselectivity is observed with an increase in temperature, an optimum being reached in the range 30–50 °C. This phenomenon has been explained by the existence of a monomer/dimer equilibrium for structure **40**.

Problem

12 → **13** + **14**

1. $MeAlCl_2$, CH_2Cl_2, -78 °C, 41%
2. PhCOCl, DMAP, Py, 95%

Tips

- The first reaction is a cationic cyclization.
- Cyclization is initiated by attack from a Lewis acid.
- The second step accomplishes an esterification.
- The epoxide is the most nucleophilic site in the molecule.
- Attack by Lewis acid leads to the thermodynamically most stable cation.
- In both products the configurations of the five newly formed stereocenters are identical. The two differ from each other only in the position of the double bond.
- The correct configuration of the stereocenters in products **13** and **14** can be deduced from the illustrated reactive conformation for structure **12**.

Solution

Methylaluminum dichloride opens the epoxide regioselectively to a tertiary carbocation. This initiates a stereoselective cyclization cascade to the pentacyclic compounds **48** and **49**, in the course of which five stereocenters are created. Compounds **48** and **49** are formed in a ratio of 3:2. After esterification with benzoyl chloride, **13** and **14** are separable chromatographically. Isomer **13** can be transformed into **14** by simple heating in HCl/acetic acid. Recrystallization produces pentacyclic substance **14** in enantiomerically pure form. Selective construction of the stereocenters can be understood graphically by imagining reaction occurring from the chair-boat-chair conformation shown in **47**. Further prerequisites are the 1–5 separation of the various double bonds and the formation of tertiary carbocations after each ring closure.

Discussion

The principle underlying the ring-closure reaction and the structure of **47** is reminiscent of the natural biosynthetic pathway for steroids.[16] In the latter case 2,3-oxidosqualene cyclizes under enzymatic control to lanosterol, a tetracyclic precursor of cholesterol.

In another biosynthetic analogy, *Johnson*[17] has synthesized an acyclic compound that cyclizes under mild conditions to the tetracyclic steroid skeleton. More recent work has examined the mechanism of this cyclization reaction.[18] Syntheses suggested by natural biosynthetic routes are referred to as "biomimetic syntheses."

Problem

Tips

- An organometallic reagent, described as a carbenoid species, is capable of serving as a methylene source.
- Addition of the carbenoid occurs regioselectively at one of the two double bonds.
- Cyclopropanation takes place stereoselectively from the sterically less demanding side of the molecule.

Solution

Stereoselective and stereospecific addition of the zinc carbenoid species ICH_2-Zn-I (**50**) occurs at the more electron-rich double bond at position 17.[19]

Problem

1.
2.
3. $PhCO_3{}^tBu$, CuBr (cat.), PhCl, 115 °C
4.

61%

15 → **16**

Tips

- In the first two transformations an ester is converted into a silyl ether.
- The third reaction is an oxidative radical process, in the course of which a ring is opened and a benzoate ester is created.
- The perbenzoate ester decomposes under copper(I) bromide catalysis into a benzoate anion and a *tert*-butoxy radical. This radical abstracts a hydrogen atom at an allylic position.
- A methyl radical recombines with the benzoate anion with regeneration of copper(I) bromide.
- The final transformation includes release of a primary alcohol.

Solution

Reduction of the ester group to give secondary alcohol **54** and subsequent protection with *tert*-butyldimethylsilyl chloride leads to silyl ether **55**. Oxidative cleavage of the cyclopropyl ring in this system proceeds through a radical mechanism and results in benzoate ester **56**. Basic ester hydrolysis gives primary alcohol **16**.

Perbenzoic acid *tert*-butyl ester (**51**) is the source – under copper(I) bromide catalysis – of a benzoate anion (**52**) and radical **53**. Radical **53** subsequently abstracts a hydrogen atom selectively from the 11-position of **55** in a homolytic bond cleavage to give a butadiene system with opening of the cyclopropane ring.[20]

$PhCO_3{}^tBu$ (**51**) —CuBr→ $PhCO_2^{\ominus}$ (**52**) + $^{\oplus}Cu(II)Br$ + • O^tBu (**53**)

$PhCO_2^{\ominus}$ $^{\oplus}Cu(II)Br$

- CuBr

- HO^tBu

54: R = H
55: R = TBS

56: R = TBS

Recombination of the intermediate methyl radical (at position 17) with **52** provides benzoate ester **56** and regeneration of the copper(I) bromide. This reaction is also known as a *Kharasch* oxidation.[19]

1. DIBAH, CH_2Cl_2.
2. TBSCl, imidazole, DMF, 55 °C, 95% over two steps.
3. Aq. NaOH, MeOH/THF, 96%.

Problem

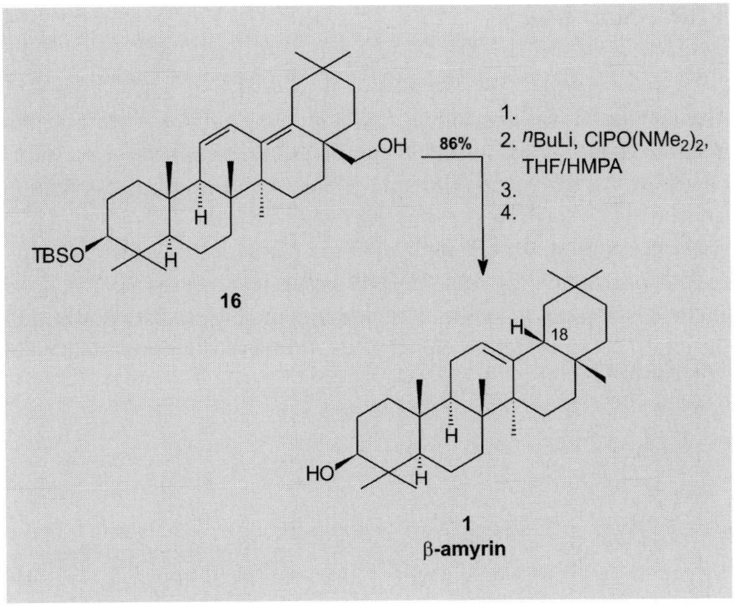

1.
2. ^nBuLi, $ClPO(NMe_2)_2$, THF/HMPA
3.
4.

86%

16

1
β-amyrin

Tips

- The reduction process used in the first and third reactions has already been encountered in this chapter.
- A new stereocenter is created during the first transformation.
- The proton at position 18 is introduced through an intramolecular reaction.
- In the third step the methyl group is generated. A phosphoric acid diamide created in the third reaction is cleaved under reductive conditions.
- Finally, use of a halide ion permits the generation of β-amyrin (**1**).

Solution

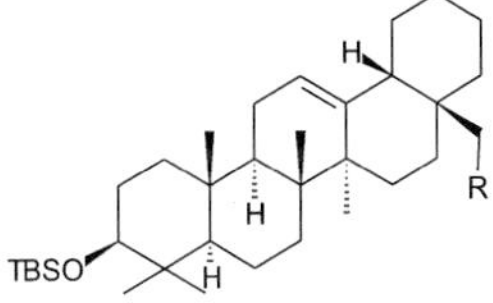

Stereoselective 1,4-reduction of the 1,3-butadiene system to olefin **57** takes place under the conditions of the *Birch* reduction. Intramolecular protonation of the intermediate carbanion at the 18-position to give **57** occurs with high selectivity *syn* to the hydroxymethylene group. Conversion into phosphoric acid derivative **58** and cleavage of the phosphoric acid amide group under the conditions of the *Benkeser* reduction provides compound **59**.[21] Fluoride ion causes the release of free β-amyrin (**1**) in a final step.

1. Li, NH_3(l)/THF (1/1.75), –78 °C, 93%.
2. ^nBuLi, $ClPO(NMe_2)_2$, THF/HMPA, 86%.
3. Li, $NEtH_2$, ^tBuOH, THF, 0 °C, 84%.
4. $^nBu_4NF$, THF, 55 °C, 95%.

57: R = OH
58: R = $OPO(NMe_2)_2$
59: R = H

12.4 Summary

The synthesis of β-amyrin (**1**) discussed here provided the first enantioselective access to the naturally occurring pentacyclic triterpenes. In a single step it proved possible with a cationic cyclization to generate five different stereocenters selectively.

From a retrosynthetic point of view, pentacyclic system **1** is dissected in a straightforward way via structure **12** to iodide **7** and bicyclic compound **6**, both of which are readily accessible starting materials. β-Amyrin was prepared in 27 steps with an overall yield of 0.013%.

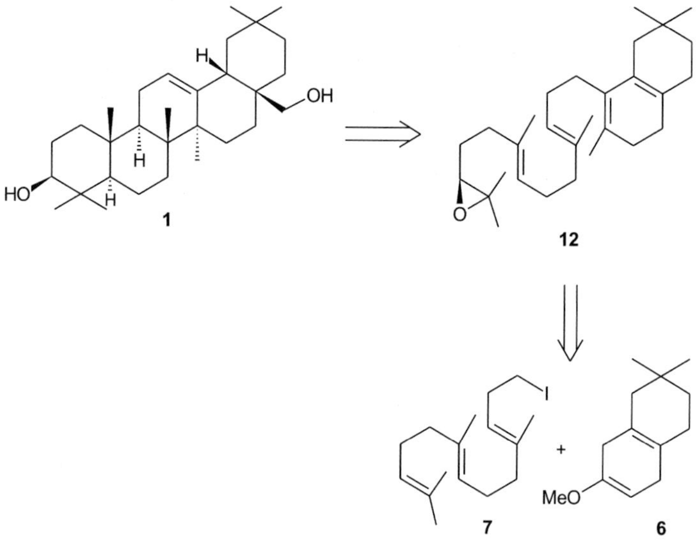

12.5 References

1 K.V. Rao, P.K. Bose, *J. Org. Chem.* **1962**, *27*, 1470; C.R. Noller, J.F. Carson, *J. Am. Chem. Soc.* **1941**, *63*, 2238.

2 E.J. Corey, H.J. Hess, S. Proskow, *J. Am. Chem. Soc.* **1963**, *85*, 3979.

3 D.H.R. Barton, E.F. Lier, J.F. McGhie, *J. Chem. Soc. (C)* **1968**, 1031.

4 E.E. van Tamelen, M.P. Seiler, W. Wierenga, *J. Am. Chem. Soc.* **1972**, *85*, 8229.

5 R.E. Ireland, D.M. Walba, *Tetrahedron Lett.* **1976**, 1071.

6 T. Kametani, Y. Hirai, Y. Shiratori, K. Fukumoto, S. Satoh, *J. Am. Chem. Soc.* **1978**, *100*, 554.

7 W.S. Johnson, M.S. Plummer, S. Pulla Reddy, W.R. Bartlett, *J. Am. Chem. Soc.* **1993**, *115*, 515.

8 A.J. Birch, H. Smith, *Q. Rev. Chem. Soc.* **1958**, *7*, 17; D. Caine, *Org. React.* **1976**, *23*, 1; J.M. Hook, L.N. Mander, *Natural Prod. Rep.* **1986**, *3*, 35; P.W. Rabideau, *Tetrahedron* **1989**, *45*, 1597.

9 E.J. Corey, M. Jautelat, *Tetrahedron Lett.* **1968**, 5787.

10 J.B. Hendrickson, R. Bergeron, *Tetrahedron Lett.* **1973**, 4607; J.E. McMurry, W.J. Scott, *Tetrahedron Lett.* **1983**, *24*, 979.

11 H. Gilman, R.G. Jones, L.A. Woods, *J. Org. Chem.* **1952**, *17*, 1630.

12 J.E. McMurry, W.J. Scott, *Tetrahedron Lett.* **1980**, *21*, 4313.

13 J.A. Mancuso, D. Swern, *Synthesis* **1981**, 165.

14 H.M. Sirat, E.J. Thomas, J.D. Wallis, *J. Chem. Soc., Perkin Trans. I* **1982**, *1*, 2885.

15 E.J. Corey, K.Y. Yi, S.P.T. Matsuda, *Tetrahedron Lett.* **1992**, *33*, 2319 und 4141; E.J. Corey, R.K. Bakshi, S. Shibata, *J. Am. Chem. Soc.* **1987**, *109*, 5551; review: S. Wallbaum, J. Martens, *Tetrahedron: Asymmetry* **1992**, *3*, 1475.

16 D. Voet, J.G. Voet in *Biochemistry*, (Ed.: A. Maelicke, W. Müller-Esterl), Wiley-VCH, Weinheim 1992.

17 W.S. Johnson, *Bioorganic Chemistry* **1978**, *5*, 51; W.S. Johnson, S.J. Telfer, S. Cheng, V. Schubert, *J. Am. Chem. Soc.* **1987**, *109*, 2517 and W.S. Johnson, S.D. Lindell, J. Steele, 5852; D. Guay, W.S. Johnson, U. Schubert, *J. Org. Chem.* **1989**, *54*, 4731.

18 E.J. Corey, H.B. Wood, *J. Am. Chem. Soc.* **1996**, *118*, 11982; E.J. Corey, S.C. Virgil, H. Cheng, C.H. Baker, S.P.T. Matsuda, V. Singh, S. Sarshar, *J. Am. Chem. Soc.* **1995**, *117*, 11819.

19 J. Furukawa, N. Kawabata, J. Nishimura, *Tetrahedron* **1968**, *24*, 53; S.E. Denmark, J.P. Edwards, *J. Org. Chem.* **1991**, *56*, 6974.

20 M.S. Kharasch, G. Sosnovsky, N.C. Yang, *J. Am. Chem. Soc.* **1959**, *81*, 5819; H.L. Goering, U. Mayer, *J. Am. Chem. Soc.* **1964**, *86*, 3753.

21 R.A. Benkeser, J. Kang, *J. Org. Chem.* **1979**, *44*, 3737; R.A. Benkeser, E.M. Kaiser, *Org. Synth. Coll. Vol. 6*, **1988**, 852.

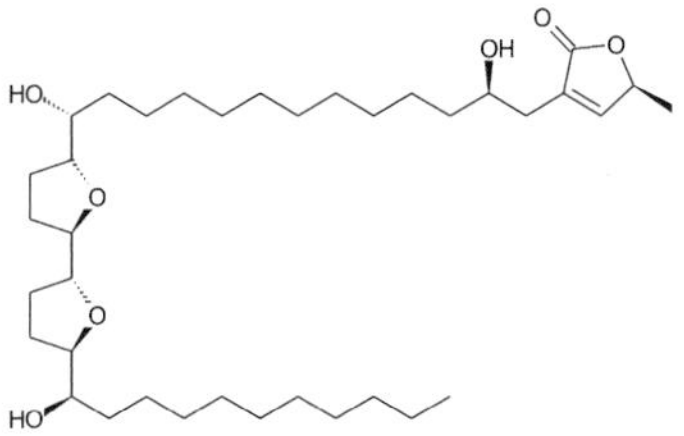

13

(+)-Asimicin: Hoye (1995)

13.1 Introduction

Asimicin belongs to the family of acetogenins, which are distinguished by their high cytotoxicity.[1] This astonishing degree of activity is explained by the fact that acetogenins inhibit mitochondrial respiration by blocking complex I (NADH/ubiquinone reductase) and thus suppressing oxidative phosphorylation required for the generation of ATP. It is further thought that the acetogenins might be valuable in combination therapy involving other cytostatic agents, because many multiresistant germs have developed ATP-dependent transport systems for eliminating cell toxins (e.g., cytostatic agents) from the cell.

The acetogenins are distinguished by considerable structural variety.[1] They all display at least one tetrahydrofuran unit, a hydroxy group, and a γ-lactone, but one also encounters epoxides, double bonds, and ketone moieties, with variable locations for the several substituents.

Investigations into structure-activity relationships provide the following picture: Compounds with two tetrahydrofuran units are more active than those with only one, especially if the two are adjacent. Biological activity reaches a maximum at three hydroxy groups and diminishes with oxidation or acetylation. One other decisive structural factor is the unsaturated γ-lactone ring, together with the alcohol group at C-4; both reduction of the double bond and translactonization lead to a reduction in activity.

The synthesis by *Hoye*[2] presented here takes advantage of multiple domino reactions to assemble a structurally complex molecule in relatively few steps.

13.2 Overview

EtO2C
C10H21
2
1.
2.
O
O
OH
O
O
C10H21
3
1.
2.
O
OMe
O
O
O
C10H21
4
O
O
O
HO
C10H21
5
1. TBSCl
2. DIBAH, -78 °C
3. PPh3P=CHCO2Et
4. DIBAH, RT
66% over 4 steps
6
OH
HO
O
O
TBSO
C10H21
7
1. TPSCl, 86%
2.
3.TBAF 8.0 eq, 88%
O
O
O
HO
C10H21
8
1.
2.
HO
O
O
HO
C10H21
9

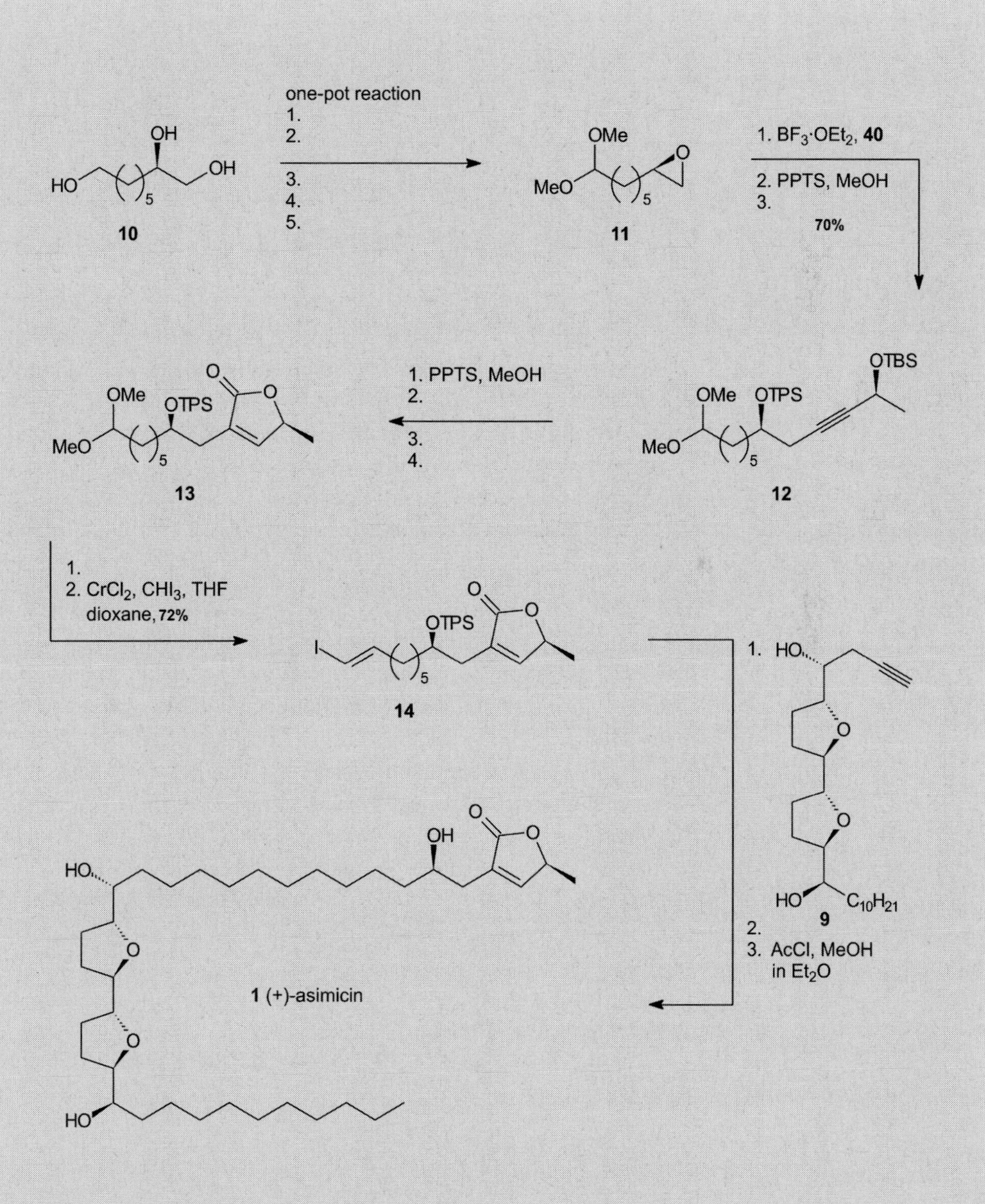

one-pot reaction
1.
2.
3.
4.
5.
10
11
1. $BF_3·OEt_2$, 40
2. PPTS, MeOH
3.
70%
12
1. PPTS, MeOH
2.
3.
4.
13
1.
2. $CrCl_2$, CHI_3, THF dioxane, 72%
14
1.
9
2.
3. AcCl, MeOH in Et_2O
1 (+)-asimicin

13.3 Synthesis

Problem

EtO$_2$C

1.

2.

O

O

OH

O

O

C$_{10}$H$_{21}$

2

C$_{10}$H$_{21}$

3

Tips

- A comparison of the stereochemistry of the product with that of the starting material provides a clue to the reaction sequence.
- Formation of the acetal occurs in the final step.
- In the course of this reaction sequence, alkene groups are converted into enantiomerically pure *cis*-diols.
- *Sharpless* dihydroxylation takes place under basic conditions, so another reaction occurs after formation of the two diols in situ.

Solution

MeO OMe

15

The *E,E*-diene **2** is transformed into a crystalline triol lactone with the use of AD-Mix-β. *Sharpless* dihydroxylation is carried out under basic conditions, so transesterification to a lactone follows automatically in a domino reaction.[3] Transacetalization with acetone dimethylacetal **15** under acidic catalysis provides acetal **3**. The double dihydroxylation procedure[4] leads in a single step to four stereogenic centers. Stereochemical examination of **3** makes it apparent that even though the starting material is an *E*-alkene, the product is formally a *trans* diol given the way the structure is written.

Entropy is the driving force for transacetalization, and equilibrium is shifted further toward the product by the use of acetone as solvent.

1. AD-Mix-β (K_2CO_3, $K_2OsO_2(OH)_4$, $K_3Fe(CN)_6$, $[(DHQD)_2$ PHAL] in ^{t}BuOH/H_2O).
2. Acetone dimethylacetal **15**, acetone, PTSA, 72% over the two steps.

Problem

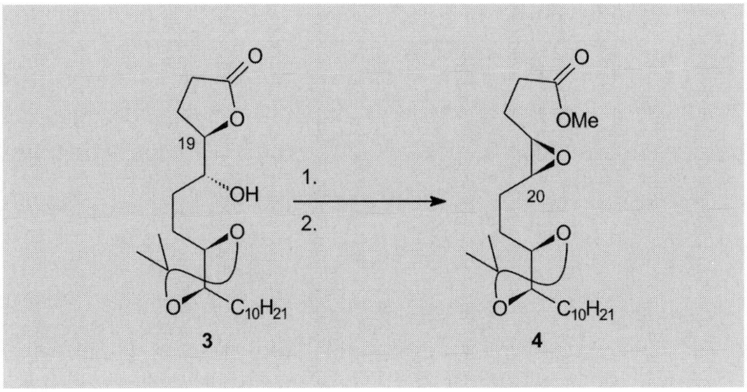

Tips

- The stereochemistry of the epoxide suggests something about the mechanism of its formation.
- The first step is a transformation of the free hydroxy group at C-20.
- A leaving group is created.
- What reaction conditions should lead to the formation of both an ester and an epoxide if a tosylate group were present at C-20?
- Methanolation opens the lactone to a methyl ester.
- The resulting alkoxide at C-19 reacts with the tosylate through intramolecular nucleophilic substitution.
- Formation of **4** occurs in a domino reaction.

Solution

The stereochemistry of epoxide **4** reveals that it has arisen through an intramolecular S_N2 reaction. Therefore, in the first step the free hydroxy group at C-20 must be converted into a tosylate. Attack on the leaving group requires that the lactone be cleaved in such a way that an alkoxide is formed at C-19. Attack by a methoxide anion transforms the lactone into a methyl ester, generating an alkoxide at C-19. The presence of sufficient methoxide anion can be achieved under even weakly basic conditions (potassium carbonate in methanol).

This reaction cascade is once again a domino reaction, because conversion of the lactone into an ester is what makes it possible for the epoxide to form.

1. Tosyl chloride.
2. K_2CO_3, MeOH, 91% over two steps.

Problem

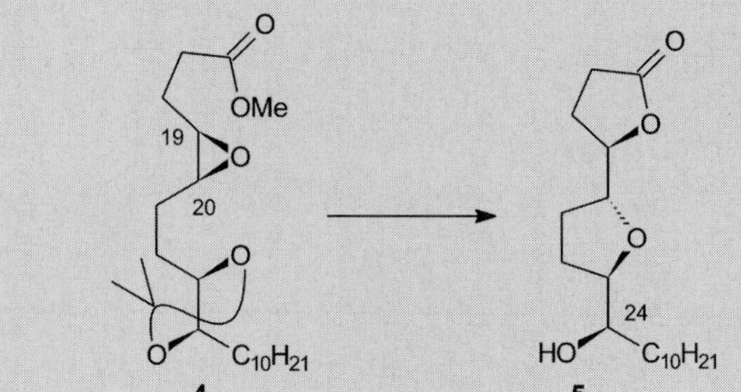

Tips

- Here again, comparison of the stereochemistries of product and starting material provides insight into the course of the reaction.
- This is yet another domino reaction.
- The configuration at C-20 is inverted because the reaction is an intramolecular nucleophilic substitution.
- The reaction sequence is initiated at the acetal function.
- Acetals are stable to base.
- The same reagent simultaneously activates both the epoxide and the ester for subsequent reactions.

Solution

The reagent utilized here is $BF_3 \cdot OEt_2$. A possible mechanism for the formation of **5** has $BF_3 \cdot OEt_2$ first acting as a Lewis acid to cleave the acetal to enol ether **16** and then activating the epoxide for intramolecular nucleophilic attack by the alkoxide at C-20. In this way the stereochemistry of the tetrahydrofuran ring in **17** is established.

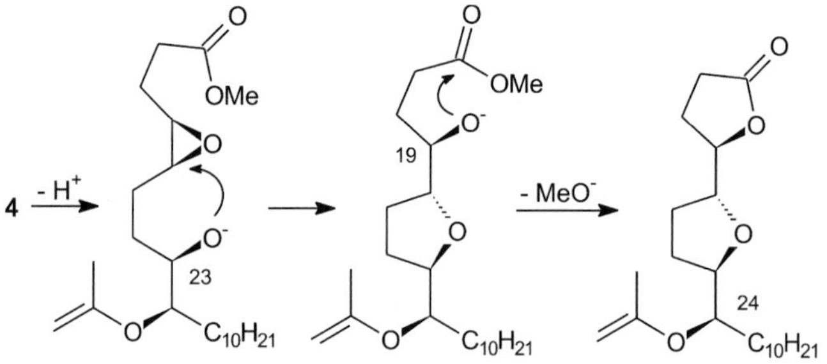

The alkoxide anion at C-19 that results from opening the epoxide reacts to form lactone **18**, another transesterification catalyzed by the Lewis acid. Aqueous workup cleaves enol ether **18** at C-24 to give free alcohol **5**.
20 mol% $BF_3 \cdot OEt_2$, CH_2Cl_2, 67%.

Problem

1. TBSCl
2. DIBAH, -78°C
3. $PPh_3P{=}CHCO_2Et$
4. DIBAH, RT
66%

5 **6**

Tips

- The first step is a protecting-group operation.
- DIBAH at low temperature reduces the lactone only as far as a lactol.
- The lactol is present as an equilibrium mixture of two structures.
- Only one form of the lactol is able to react with a *Wittig* ylide.
- The reaction with the *Wittig* ylide takes place under neutral conditions without the need for additional base.
- Stabilized ylides lead to only a single configuration about the double bond.
- The ester is reduced with DIBAH in the final step to give an allylic alcohol.

Solution

In the first step the free hydroxy group is protected as a *tert*-butyldimethylsilyl ether. Then the lactone is reduced to a lactol, which is the cyclic hemiacetal of an aldehyde. Overreduction to an alcohol can be prevented by stoichiometric addition of the reducing reagent at low temperature. Selectivity here depends on the relative stability of intermediate **19**, which decomposes only in the course of work-up[5] (see Chapter 3).

The hemiacetal exists as an equilibrium mixture of cyclic compound **20** and its open counterpart (**21**), but an aldehyde addition reaction can occur only with the acyclic form. A *Wittig* reaction of the stabilized ylide leads to an α,β-unsaturated ester that has the *E* configuration with respect to the double bond. This reaction occurs under neutral conditions, so 1,4-addition of the alcohol to the α,β-unsaturated ester is avoided.[6] A subsequent DIBAH reduction leads to allylic alcohol **6** in a reaction that ordinarily shows complete 1,2-selectivity.

19

20 **21**

Problem

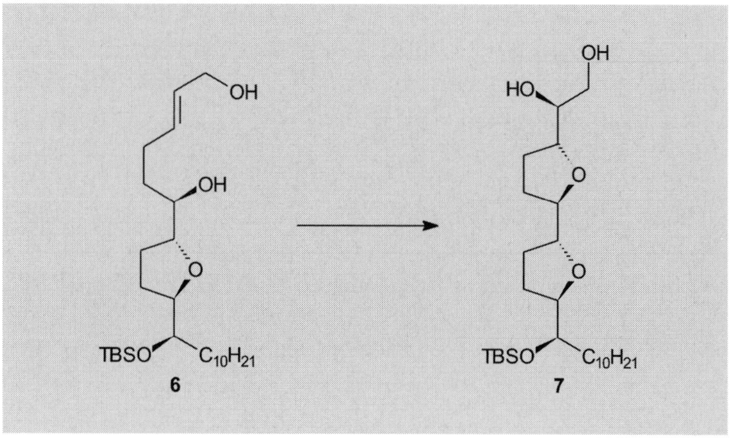

Tips

- Which functional groups and bonds are new?
- Allylic alcohols are specific substrates for a particular asymmetric oxidation method.
- The two enantiomeric forms of the diethyl tartrate ligand in *Sharpless* oxidation lead to epoxy alcohols **22** and **24**. Which of the isomeric ligands is required for the present reaction?

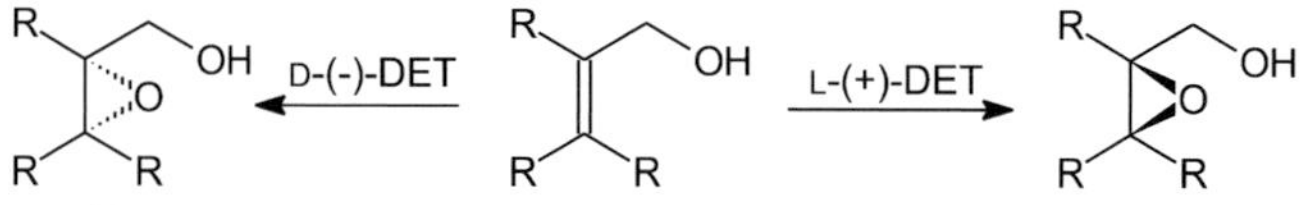

- One more domino reaction has taken place!

Solution

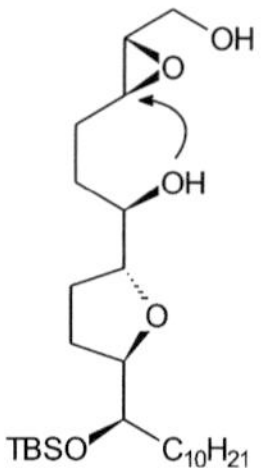

The allylic alcohol is transformed through a *Sharpless* epoxidation[7] with (+)-DET, ^{t}BuOOH, and Ti(O^{i}Pr)$_4$ into the corresponding epoxy alcohol **25** (see Chapter 6), which is further reacted to diol **7** by subsequent attack of the alcohol on the epoxide with formation of a second tetrahydrofuran ring. In this case it proved necessary as a result of several factors to use 50 mol% of the titanium reagent. This is because: 1) The other free hydroxy group in the substrate competes with the allylic alcohol for the catalyst; 2) The product diol binds the titanium reagent in a chelate complex; and 3) Titanium acts also as a Lewis acid for activating the epoxide.

(+)-DET, ^{t}BuOOH, 50 mol% Ti(O^{i}Pr)$_4$, 87% yield with 50% conversion.

Problem

Tips

- Formation of the epoxide occurs intramolecularly.
- Is it the primary or the secondary alcohol that is transformed into a leaving group?
- In the third step the epoxide arises through a domino reaction.

Solution

The first step causes the primary alcohol to be converted selectively into the corresponding *tert*-butyldiphenylsilyl ether. It is possible to differentiate between the primary and secondary hydroxy functions because primary alcohols react considerably more rapidly for steric reasons. The stereochemistry at C-15 is inverted, so the epoxide must have been prepared in an intramolecular S_N2 reaction in which the leaving group was located on the secondary alcohol. For this reason the latter esterified to the tosylate (**26**). The silyl protecting group is cleaved with TBAF in THF. Here again, use is made of a domino reaction, because upon cleavage of the silyl residue, alkoxide **27** reacts in a nucleophilic substitution reaction to give an epoxide.

Problem

Tips

- With which reagent combination is the triple bond best introduced: a) ^{n}BuLi, acetylene; b) $Me_3Sn\text{-}C{\equiv}C\text{-}SiMe_3$, TBAF; c) CuI, acetylene; or d) $Li\text{-}C{\equiv}C\text{-}SMe_3$, $BF_3 \cdot OEt_2$?
- What is the minimum required number of equivalents of the reagent $Li\text{-}C{\equiv}C\text{-}SMe_3$: a) 1.0; b) 2.0; or c) 4.0?
- $BF_3 \cdot OEt_2$ increases the reactivity of the epoxide.
- With which reagent combination can a silyl group be cleaved from an alkyne: a) TBAF; b) K_2CO_3, MeOH; or c) $AgNO_3$, KCN?

Solution

Since in compound **8** the alcohol group at C-24 is unprotected, 2.8 equivalents of the reagent $Li\text{-}C{\equiv}C\text{-}SMe_3$ are required. $BF_3 \cdot OEt_2$ is again employed as a Lewis acid as a way of increasing the reactivity of the epoxide.[8] The silyl group is cleaved with methoxide (**3** → **4**).

1. 2.8 eq. $Li\text{-}C{\equiv}C\text{-}SMe_3$, $BF_3 \cdot OEt_2$.
2. K_2CO_3, MeOH, 70% over two steps.

Discussion

Trimethylsilylacetylene (**29**) is preferable to ethyne for several reasons: 1) It is a liquid rather than a gas; 2) The metallation is more easily controlled; and 3) Bisadducts are avoided, because only one reactive site is present.

All the reagents listed can be used to cleave trimethylsilyl groups from acetylenes: fluoride, potassium carbonate under basic conditions in methanol, or silver nitrate/potassium cyanide.[9] With the third method advantage can be taken of the fact that the later transition metals (e.g., copper or silver) complex readily with acetylides. Workup with concentrated potassium cyanide solution causes compound **32** to be cleaved to alkyne **33**. In this way silylated alkynes can be deprotected in the presence of *O*-silyl groups.

$R-\equiv-SiMe_3$ (30) $\xrightarrow{AgNO_3}$ $R-\equiv-SiMe_3$ with Ag^+, H_2O (31) $\xrightarrow{-Me_3SiOH}$

$R-\equiv-Ag$ (32) $\xrightarrow[H_2O]{KCN}$ $R-\equiv-H$ + AgCN (33)

This completes the synthesis of the first building block. The synthesis of the second is discussed below.

Starting material **10** for the second building block is obtained from 1,7-octadiene by monohydroboration and subsequent dihydroxylation. After recrystallization, enantiomerically pure **10** is isolated in 64% yield over this complete set of steps.

Problem

10 → one-pot reaction 1. 2. 3. 4. 5. → **11**

Tips

- The three formal steps required for synthesis of **11** are oxidation, creation of a dimethylacetal, and epoxide formation. They are carried out in this order as steps three through five.
- Look closely at the stereochemistry of the epoxide!
- With what reagents can the two primary alcohol groups be distinguished? Possibilities include: a) tosyl chloride; b) *tert*-butyldimethylsilyl chloride; c) Ac_2O; and d) trimethyl orthoacetate, PPTS.
- The differentiation is accomplished by reaction of the 1,2-diol structure with trimethyl orthoacetate, in the course of which a cyclic intermediate is formed.
- 2-Methoxy-1,3-dioxolane is activated with TMSCl. The electrophilic portion of TMSCl reacts as a Lewis acid, and the nucleophilic portion appears in the product.
- The dioxolane group becomes an acetate.
- In the third and fourth steps a dimethylacetal is constructed from the remaining primary alcohol function.
- This is a reaction sequence consisting of an oxidation and an acetalization.
- Synthesis of the epoxide proceeds from a 1-chloro-2-acetoxy compound.

2-methoxy-1,3-dioxolane

Solution

Conducting the first two steps as a one-pot reaction[10] was first suggested by *Sharpless*. Transacetalization of trimethyl orthoacetate with **10** under weakly acidic catalysis leads to the 2-methoxy-1,3-dioxolane **34**. Addition of TMSCl or acetyl chloride generates the 1,3-dioxolanylium cation **35**, which is opened nucleophilically and regioselectively at the sterically least hindered position to chloroacetate **36**.

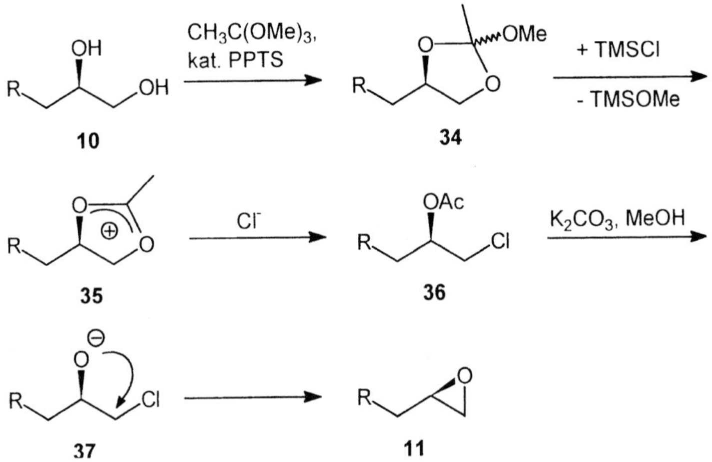

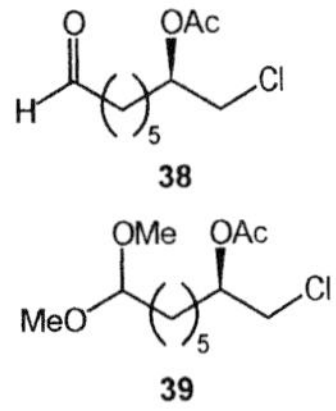

Hoye extended this convenient reaction technique to include the subsequent steps as well. After the first two steps, all volatile components are removed and the reagents for a *Swern* oxidation[11] are added. Then all volatile materials are again removed. Once acetalization of **38** with methanolic PTSA to give **39** is complete, K_2CO_3 is added in excess. The reaction mixture is thereby neutralized, after which the acetate is saponified. The resulting alkoxide **37** attacks the chlorine-containing carbon in an S_N2 reaction leading to epoxide **11** with retention of configuration.

1. $CH_3C(OMe)_3$, PPTS, CH_2Cl_2.
2. TMSCl.
3. DMSO, $(COCl)_2$, NEt_3.
4. PTSA, MeOH.
5. K_2CO_3, MeOH, 86% over five steps.

Problem

Tips

- The first step is analogous to the transformation **8** → **9**.
- $BF_3 \cdot OEt_2$ activates the epoxide. Side reactions may occur.
- The C_4-unit is introduced in the form of a lithiated alkyne.
- In the second step one of the functional groups must be reprotected.
- The third step is also a protecting-group operation.
- A *tert*-butyldiphenylsilyl ether is prepared.

Solution

The epoxide is activated with $BF_3 \cdot OEt_2$ and opened with the protected butynol **40**.[12] Since the Lewis acid present to some extent cleaves the acetal, the aldehyde must once again be protected. The alcohol function is subsequently transformed into the corresponding *tert*-butyldiphenylsilyl ether **12**.

1. $BF_3 \cdot OEt_2$, **40**.
2. PPTS, methanol.
3. TPSCl.

Problem

Tips

- What functional groups have been altered?
- The first step is a protecting-group operation.
- The second step involves reduction of a propargylic alcohol to an *E*-allylic alcohol.
- Which reagent would give an *E*-alcohol: a) *Lindlar* catalyst, H_2; or b) Red-Al®?

- The intermediate in reduction with Red-Al® is functionalized using iodine as an electrophile.
- The missing carbon atom is introduced in the form of carbon monoxide.
- The vinylic iodide is extended by one carbon with carbon monoxide.
- This is a palladium-catalyzed reaction.

Solution

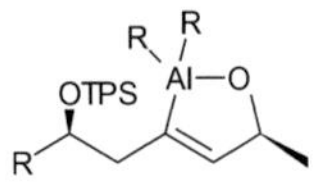

The TBS protecting group is selectively cleaved in the first step. The released propargylic alcohol is then reduced with Red-Al® to an allylic alcohol. Reduction of the propargylic alcohol occurs selectively *trans*. Since during this reaction a five-membered ring system **41** is created through coordination of aluminum with the alkoxide, it is assured that aluminum is located in the β-position relative to the alkoxide.

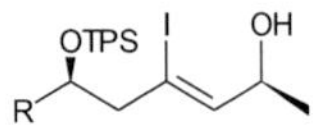

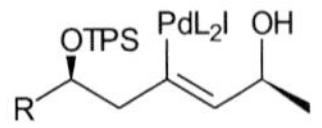

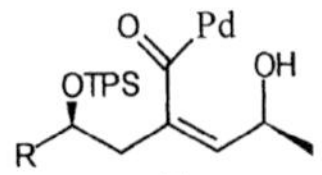

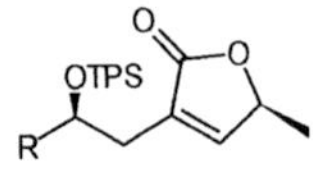

The carbanion is trapped with iodine to give **42**, which makes a further functionalization possible. Conversion of vinylic iodide **42** into a lactone is accomplished by palladium-catalyzed carbonylation under *Stille* conditions.[13] This process can be broken down into the following elementary reactions: a) Oxidative addition of Pd^0 to vinylic iodide **42** with formation of **43**; b) An insertion reaction of carbon monoxide with creation of the pallada-acyl species **44**; c) Reaction of acid-chloride equivalent **44** with the alcohol to give lactone **13**.

1. PPTS, MeOH.
2. Sodium bis(2-methoxyethoxy)aluminum hydride (Red-Al®).
3. I_2.
4. $(PPh_3)_2PdCl_2$, N_2H_4, K_2CO_3, CO (3 bar), THF, 83% over the entire group of steps.

Problem

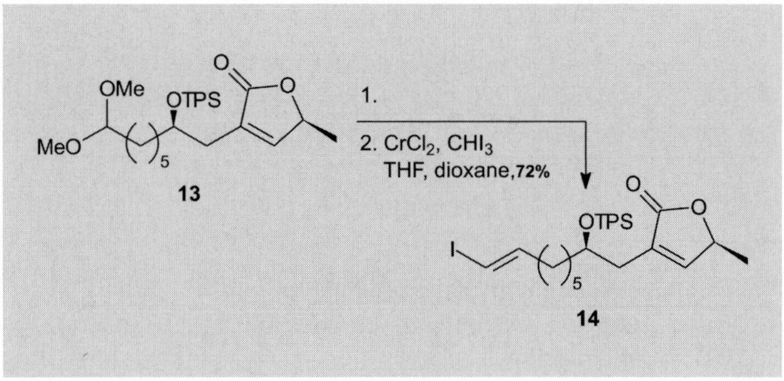

1\.

2\. $CrCl_2$, CHI_3 THF, dioxane, 72%

Tips

- The operation accomplishes a C_1 chain extension.
- The extra carbon atom is derived from CHI_3.
- The first reaction is a protecting-group operation.
- What follows involves the addition of an organometallic species to a deprotected aldehyde.
- To the aldehyde is added a reagent prepared by oxidative addition of two molecules of $CrCl_2$ to CHI_3 (see Chapter 7).
- A driving force in the reaction is the oxophilicity of chromium. Formation of an alkene occurs in a manner similar to that familiar from the *Wittig* reaction.

Solution

In the first step the acetal is cleaved to an aldehyde with aqueous trifluoroacetic acid. The subsequent C_1 chain extension was pioneered by *Takai*[14] and can be regarded as a chromium-mediated *Wittig* reaction. Two molecules of chromium(II) chloride first add oxidatively to triiodomethane with the formation of **46** (via **45**). The biscarbanion then adds to the aldehyde. Intermediate **47** next loses oxygen and two chromium atoms, probably in a concerted reaction, with creation of the double bond. From the mechanism it is apparent that an *E*/*Z* mixture of vinylic iodides will usually be obtained. In the present case the *E*/*Z* ratio was 4:1.

1. Trifluoroacetic acid, $CHCl_3$, H_2O.
2. $CrCl_2$, CHI_3, THF, dioxane, 72% over the two steps.

ICr(III)—CHI$_2$ → ICr(III), ICr(III) CHI
45
46
RCHO
Cr(III)I
R CHI
OCr(III)I
47
I R
48

Problem

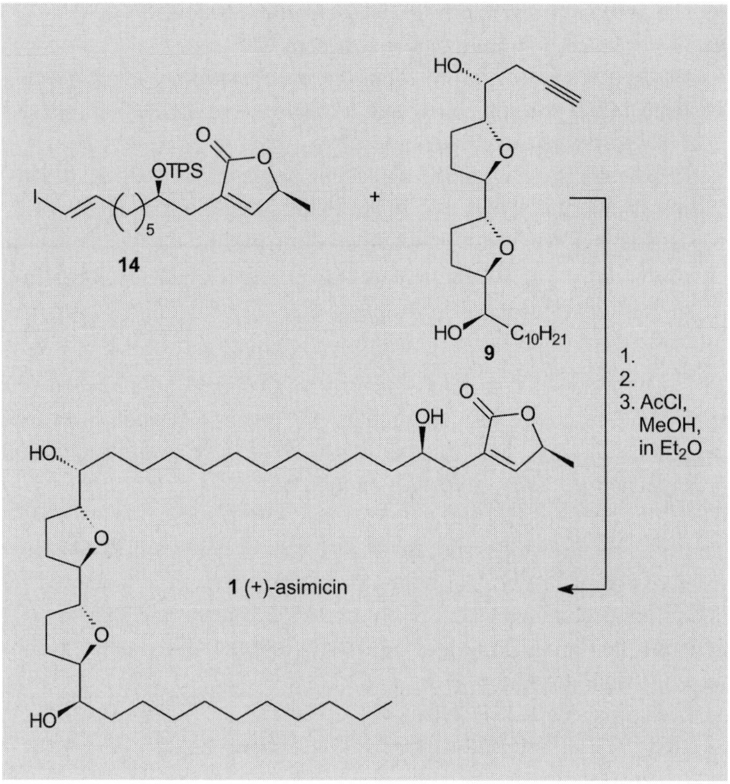

Tips

- The vinylic iodide and the alkyne are coupled in the first step.
- These two fragments are joined with a palladium-catalyzed alkyne coupling reaction.
- The product is an enyne.
- The *Wilkinson* catalyst [$Rh(PPh_3)_3Cl$] is able to distinguish between the two unsaturated functionalities in this molecule.
- Acetyl chloride and methanol react together to form the active reagent.

Solution

In the first step the two building blocks are connected with an alkyne coupling reaction introduced by *Sonogashira*.[15] *Sonogashira* coupling incorporates the following elementary reactions: 1) Oxidative addition of Pd^0 to a vinylic iodide to give complex **49**; 2) Transmetallation with a copper-*at*-copper complex to **50**; 3) Reductive elimination with release of the product and regeneration of palladium species **48**.

The catalytic cycle constituting the alkyne coupling can be represented as follows:

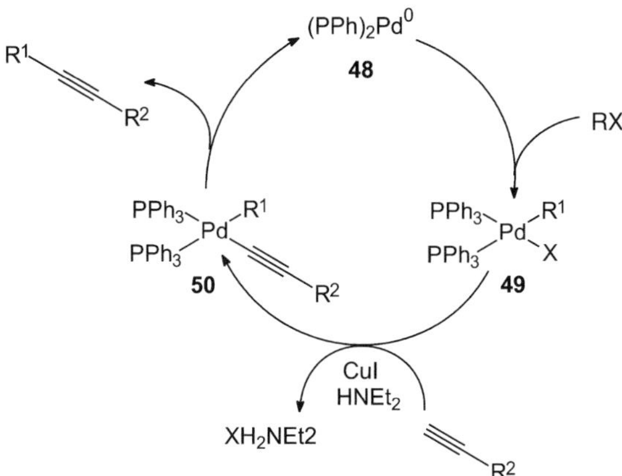

Special attention is warranted to the role of copper iodide and the unusual solvent triethylamine. Triethylamine is not a strong base, but it is nevertheless able to support a small equilibrium concentration of the alkynyl anion. Copper iodide assists in this deprotonation because the resulting *at* complex **51** is stable. The reaction occurs catalytically, so catalytic amounts of **51** are sufficient.

A side reaction leads to the formation of diynes if a second molecule of acetylide succeeds in displacing the original coupling partner from palladium, allowing reductive elimination.

The resulting enyne is then selectively hydrogenated with *Wilkinson* catalyst in benzene, in the course of which the α,β-unsaturated lactone survives untouched.

Acetyl chloride/methanol has often been employed by *Hoye* for cleaving the silyl protecting group. Mechanistically one can postulate that reaction of the acid chloride with methanol leads to an ether solution of anhydrous hydrogen chloride, which then effects the cleavage.

1. $Pd(PPh_3)_2Cl_2$, CuI, NEt_3, 79%.
2. $Rh(PPh_3)_3Cl$, H_2 (1 bar), benzene.
3. AcCl, MeOH in Et_2O, 74% over two steps.

13.4 Summary

The synthesis described is noteworthy not only for the repeated use of domino reactions but also because convergent strategy allows the number of steps to be limited to 28.

Characteristic of the strategy illustrated in this chapter is the one-pot synthesis of **11**, which was achieved in an overall yield of 86%. The reaction to give a chloroacetate permitted straightforward differentiation between the two primary hydroxy groups. In the process, the diol as a whole was subjected to transformation, and the remaining alcohol group could be selectively functionalized. Moreover, the chloroacetate represents a form of protecting group for the epoxide.

If one were to attempt to protect the diol as an acetal and then treat the remaining alcohol in the way shown, the result would be two acetals that would no longer be distinguishable chemically. The conditions for the individual partial reactions were so adapted to one another that all the transformations could be performed selectively. Alternative routes would accomplish the same result only with a considerable expenditure of effort.

The enantiomerically pure protected butynol **40** already reflects the stereochemistry of the lactone. Moreover, the chain can be extended by the missing C_4-unit in the reaction of **40** with **11**, and the alcohol at C-4 arises simultaneously in a stereoselective way. Finally, the propargylic alcohol is converted in only three steps into an α,β-unsaturated lactone.

In the first step of the conversion of **12** to **13** the *tert*-butyldimethylsilyl protecting group alone is cleaved selectively. Advantage is taken of the fact that the *tert*-butyldiphenylsilyl group is approximately 250 times more stable to acidolysis. When two protecting groups of the same type can be distinguished in such a way one speaks of "graded reactivity"[16] (see Chapter 7). Protecting groups of the same type can be separately cleaved in a similar way in the final phases of many syntheses.

The stability of the α,β-unsaturated lactone here is noteworthy, because the compound proves unreactive not only under the conditions of the *Takai* reaction but also during hydrogenation. Furthermore, *Hoye* does not mention any translactonization to the alcohol at C-4.

Joining of the two building blocks **9** and **14** was achieved with an alkyne coupling at a rather unusual point in the chain, leading to an enyne. Frequently the coupling of building blocks accompanies the incorporation of important functional groups.[17] The enyne in this case serves as the equivalent of four methylene groups.

13.5 References

1 A. Cavé, D. Cortes, B. Figadère, R. Hocquemiller, O. Laprévote, A. Laurens, M. Leboeuf, *Recent. Adv. Phytochem.* **1993**, *27*, 167; Z.-M. Gu, G.-X. Zhao, N.H. Oberlies, L. Leng, J.L. McLaughlin, *Recent. Adv. Phytochem.* **1995**, *29*, 249.

2 T.R. Hoye, L. Tan, *Tetrahedron Lett.* **1995**, *36*, 1981.

3 L.F. Tietze, *Chem. Rev.* **1996**, *96*, 115.

4 H.C. Kolb, M.S. VanNieuwenhze, K.B. Sharpless, *Chem. Rev.* **1994**, *94*, 2483; D.W. Nelson, A. Gypser, P.T. Ho, H.C. Kolb, T. Kondo, H.-L. Kwong, D.V. McGrath, A.E. Rubin, P.-O. Norrby, K.P. Gable, K.B. Sharpless, *J. Am. Chem. Soc.* **1997**, *119*, 1840.

5 F.A. Carey, R.J. Sundberg, *Advanced Organic Chemistry*, Wiley-VCH, Weinheim 1995.

6 N. Cohen, J.W. Scott, F.T. Bizzaro, R.J. Lopresti, W.F. Eichel, G. Saucy, H. Mayer, *Helv. Chim. Acta* **1978**, *61*, 837.

7 S. Masamune, W. Choy, J.S. Petersen, L.R. Sita, *Angew. Chem. Int. Ed. Engl.* **1985**, *24*, 1; T. Katsuki, K.B. Sharpless, *J. Am. Chem. Soc.* **1980**, *102*, 5974; R.A. Johnson, K.B. Sharpless, in *Catalytic Asymmetric Synthesis*, (Ed.: I. Ojima), VCH-Verlagsgesellschaft, Weinheim 1993, p. 101.

8 M. Yamaguchi, I. Hirao, *Tetrahedron Lett.* **1983**, *24*, 391.

9 H.M. Schmidt, J.F. Arens, *Rec. Trav. Chim.* **1967**, *86*, 1138.

10 H.C. Kolb, K.B. Sharpless, *Tetrahedron* **1992**, *48*, 10515.

11 A.J. Mancuso, D. Swern, *Synthesis* **1981**, 165; K. Omura, D. Swern, *Tetrahedron* **1978**, *34*, 1651; A.J. Mancuso, S.-L. Huang, D. Swern, *J. Org. Chem.* **1978**, *43*, 2480.

12 T.R. Hoye, P.E. Humpal, J.I. Jimenez, M.J. Mayer, L. Tan, Z. Ye, *Tetrahedron Lett.* **1994**, *35*, 7517.

13 A. Cowell, J.K. Stille, *J. Am. Chem. Soc.* **1980**, *102*, 4193.

14 K. Takai, K. Nitta, K. Utimoto, *J. Am. Chem. Soc.* **1986**, *108*, 7408.

15 K. Sonogashira, Y. Tohda, N. Hagihara, *Tetrahedron Lett.* **1975**, *16*, 4467; R. Rossi, A. Carpita, F. Bellina, *Org. Prep. Proc. Int.* **1995**, 27, 129.

16 M. Schelhaas, H. Waldmann, *Angew. Chem. Int. Ed. Engl.* **1996**, *35*, 2056.

17 E.M. Suh, Y. Kishi, *J. Am. Chem. Soc.* **1994**, *116*, 11205.

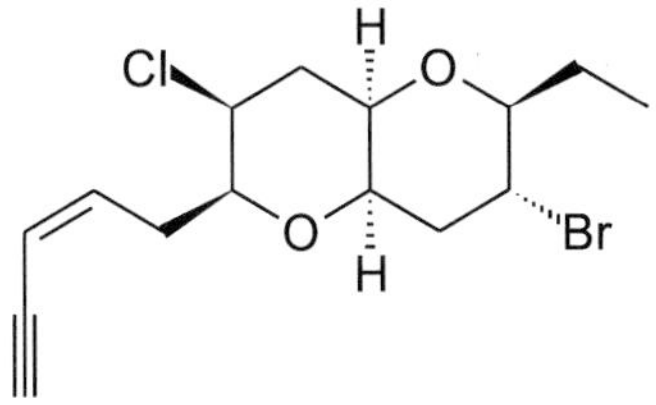

14

(Z)-Dactomelyn: Lee (1995)

14.1 Introduction

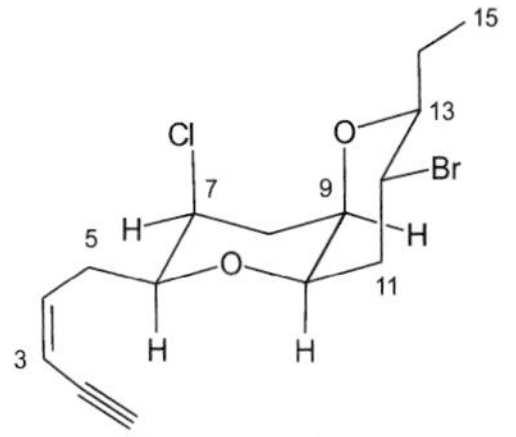

Z-Dactomelyn (**1**) was isolated by *Schmitz* in 1981 from the digestive glands of the lumpfish species *Aplysia dactylomela*.[1] Determination of the relative and absolute configurations of the six stereogenic centers was accomplished with the aid of an X-ray structural analysis of *E*-dactomelyn. The compounds are members of a group of nonisoprenoid ethers, all of which have in common a bicyclic pyran skeleton with both ethyl and pentenynyl side chains.[2] The biological function of the dactomelyns is to date unknown.

An unusual feature of dactomelyns is the presence of two halogen-substituted chiral centers. Since few methods are so far known for selective introduction of halogen atoms at sp^3-hybridized carbon, the dactomelyns pose an extraordinary synthetic challenge. Another interesting feature of these natural products is the appearance of both chlorine and bromine atoms in the same molecule. One is struck, upon closer examination of the bicyclic pyran skeleton **1**, by the fact that the chlorine atom is oriented toward the sterically less favorable side of the molecule (the concave side), whereas the bromine atom is directed toward the convex side.

First attempts on the part of *Kozikowski* to introduce halogen atoms stereoselectively into the bispyran system was unsuccessful according to a synthetic study published in 1990.[3] *Lee* successfully completed a stereoselective synthesis in 1995 by taking advantage of several radical reactions.[4]

14.2 Overview

EtO$_2$C, H, OH, HO, H, CO$_2$Et

2

(-)-diethyl tartrate

1.

2.

HO, H, OH, OH, BnO, H

3

1.

2.

3.

Cl$_3$C, H, O, Ph, O, BnO, H

4

1.

2.

Cl$_3$C, H, O, Ph, EtO$_2$C, O, O, H

5

1.2 eq ^{c}Hex$_3$SnH, 0.2 eq AIBN, C$_6$H$_6$, reflux

67%

6

1.0 eq (Me$_3$Si)$_3$SiH, 0.1 eq Et$_3$B, C$_6$H$_6$, RT

91%

Cl, H, O, Ph, EtO$_2$C, O, O, H

7

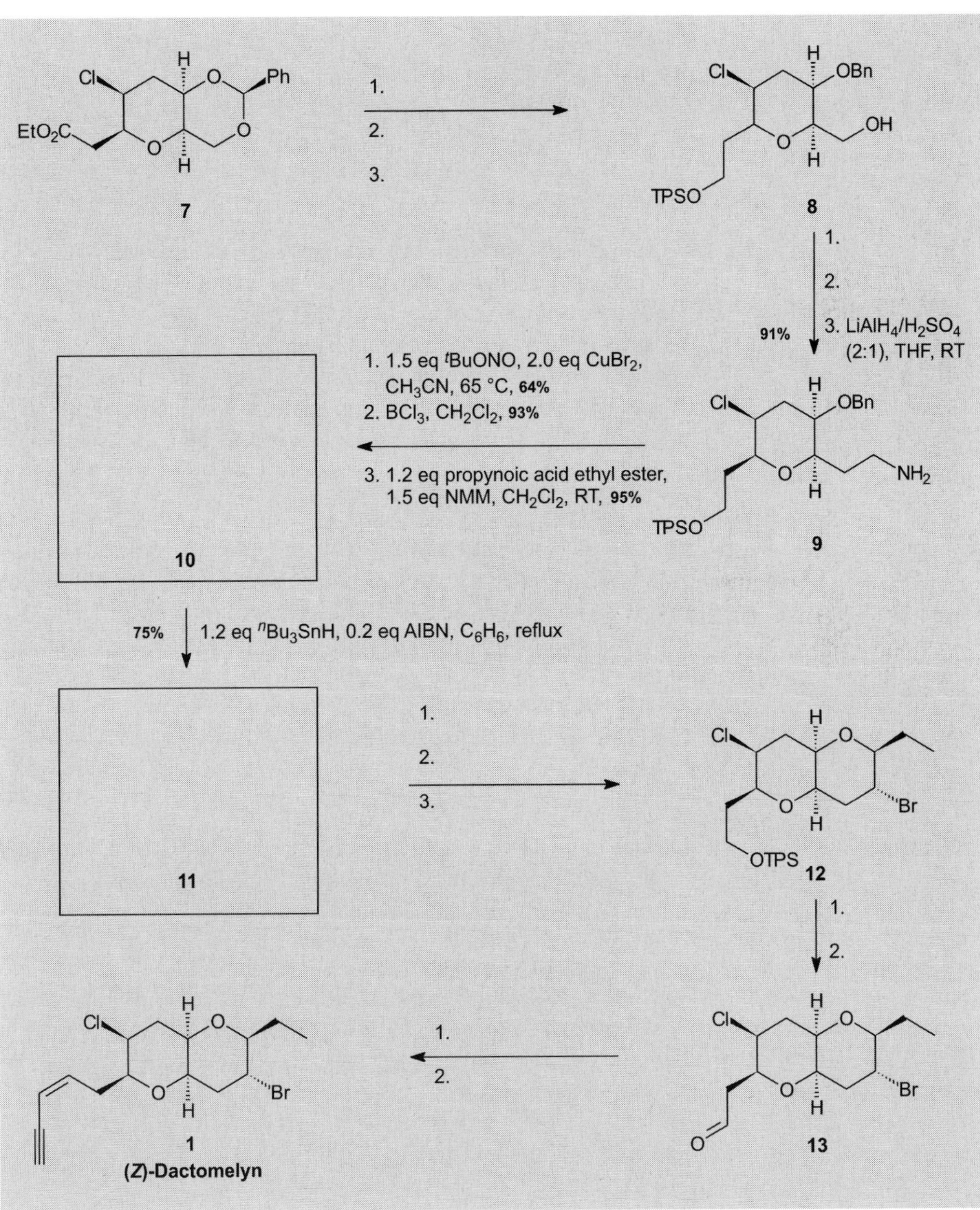

Cl
H
O
Ph
EtO2C
O
O
H
7
1.
2.
3.
H
OBn
OH
TPSO
8
1.
2.
91%
3. LiAlH4/H2SO4 (2:1), THF, RT
1. 1.5 eq tBuONO, 2.0 eq CuBr2, CH3CN, 65 °C, 64%
2. BCl3, CH2Cl2, 93%
3. 1.2 eq propynoic acid ethyl ester, 1.5 eq NMM, CH2Cl2, RT, 95%
NH2
9
10
75%
1.2 eq nBu3SnH, 0.2 eq AIBN, C6H6, reflux
11
Br
OTPS
12
13
1
(Z)-Dactomelyn

14.3 Synthesis

Problem

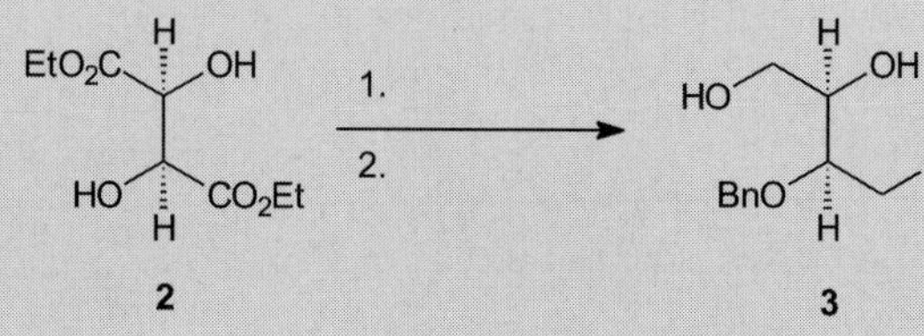

Tips

- The first step accomplishes acetalization with an aldehyde.
- The benzylic ether is derived from the acetal. What formal reaction lies behind introduction of the benzylic carbon atom: a) oxidation; b) reduction; or c) elimination?
- A 1 : 1 mixture of lithium aluminum hydride and aluminum chloride accomplishes two transformations in the second reaction. A combination of reductive and Lewis-acid characteristics make this the reagent of choice.

Solution

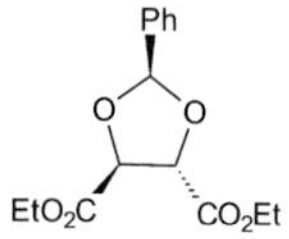

Reaction of (–)-diethyl tartrate (**2**) with benzaldehyde leads to acetal **14**. The second reaction accomplishes reduction of the ester groups to primary alcohols. Simultaneously, with the aid of a Lewis acid, the acetal is reductively cleaved to leave a benzyl protecting group.[5]

1. PhCHO, p-TsOH (cat.), C_6H_6, reflux.
2. 3.0 eq. $LiAlH_4$, 3.0 eq. $AlCl_3$, CH_2Cl_2/Et_2O, RT, 91% over two steps.

Discussion

Because of its C_2-symmetry, it is irrelevant which of the two sides of the benzylidene acetal is opened.

Problem

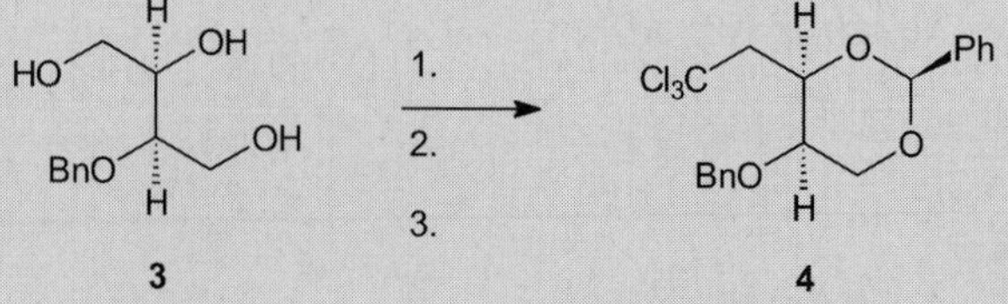

Tips

- A Cl_3C group is introduced using a simple S_N2 reaction.
- A prerequisite to this S_N2 reaction is generation of a leaving group.
- The sequence also includes introduction of a benzylidene protecting group. In what order must these transformations occur?
- A transacetalization occurs in the first step.

Solution

At this point a benzaldehyde dimethylacetal is used for the acetalization. The thermodynamically favored six-membered acetal **15** is formed preferentially over the potential competitors: five- and seven-membered acetals[5] (See Chapter 7). Transformation of a primary alcohol function into the triflate group of structure **16** must precede the S_N2 reaction in the third step, in which a carbon nucleophile is created with LDA and chloroform. Deprotonation of chloroform with LDA is carried out at –110 °C to suppress competing formation of dichlorocarbene.[6]

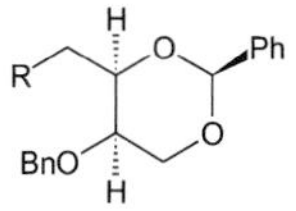

15: R = OH
16: R = OTf

1. 1.5 eq. $PhCH(OMe)_2$, *p*-TsOH (cat.), CH_2Cl_2, reflux, 83%.
2. 1.2 eq. Tf_2O, 4.0 eq. Py, CH_2Cl_2, 0 °C.
3. 0.85 eq. LDA, $CHCl_3$, THF/Et_2O/HMPA, –110 °C, 60% over two steps.

Problem

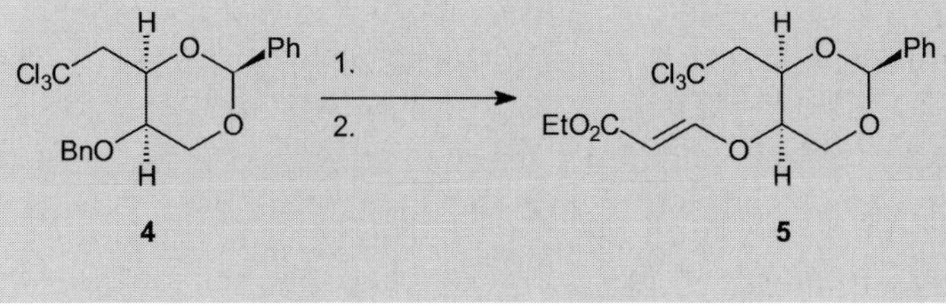

4 → (1., 2.) → **5**

Tips

- Nucleophilic addition to a triple bond serves to create the β-alkoxyacrylic ester function.
- The addition proceeds analogously to a *Michael* reaction.
- How must the alkyne be substituted in order to permit nucleophilic attack?
- The first step is a protecting-group operation.
- One protecting group is selectively cleaved under catalytic-reductive conditions.

Solution

H—≡—CO_2Et

17

Catalytic hydrogenation is a standard method for cleaving benzylic ethers. The resulting secondary alcohol constitutes the nucleophilic reaction center in a subsequent *Michael* addition to the acceptor-substituted triple bond. Compound **5** is obtained quantitatively with a double bond in the *trans* configuration.

1. Pd/C, H_2, 75%.
2. 1.2 eq. **17**, 1.5 eq. *N*-methylmorpholine, CH_2Cl_2, RT, 99%.

Discussion

A systematic investigation of the stereoselective synthesis of β-alkoxyacrylate esters was carried out as early as 1966 by *Winterfeld*.[7] It is interesting that addition of secondary amines to **17** leads exclusively to *trans*-β-dialkylaminoacrylic esters. The high *trans* selectivity in the second transformation is all the more remarkable since alcohols in the presence of a tertiary amine give addition products that are *cis*/*trans* mixtures.

Problem

Cl_3C, EtO_2C, H, O, Ph, H

5

1.2 eq ${}^{c}Hex_3SnH$, 0.2 eq AIBN, C_6H_6, reflux

67%

6

Tips

- This step is a radical cyclization.
- Where would an initially prepared Sn radical attack the molecule?
- A tin radical removes a chlorine atom.
- The cyclization passes through a chair-like transition state, the geometry of which makes it possible to predict the stereochemistry of compound **6**.
- The ester is reduced with DIBAH in a final step to give an allylic alcohol.

Solution

Trialkyltin hydrides can serve as chain carriers in radical reactions. The low Sn-H bond energy (74 kcal/mol) facilitates easy homolytic cleavage by an initiator radical to produce tin radicals that are in a position to cleave a Cl-C bond with formation of a carbon radical. The resulting carbon radical cyclizes stereoselectively to pyran **6** in a 6-*exo-trig* ring closure.[8]

The configuration of the newly formed stereogenic center can be explained conceptually by a chair-like reactive conformation for radical **18**. The β-alkoxyacrylates utilized here by *Lee* prove to be extraordinarily effective radical acceptors.[7]

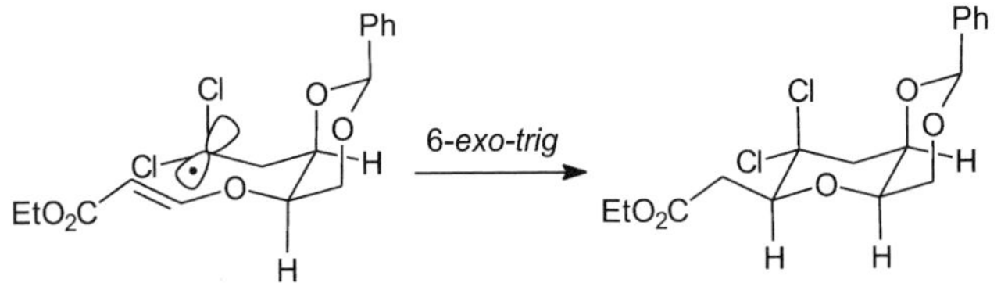

This radical reaction can be described mechanistically by the following scheme:

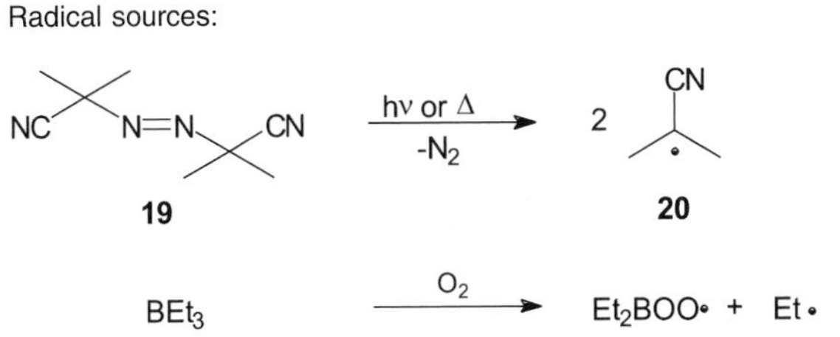

Discussion

Radical sources:

Initiator radicals can be generated in various ways. AIBN (azobisisobutyronitrile, **19**) decomposes upon exposure to light or heat into nitrogen and 2-cyanopropyl radicals (**20**). Reaction of triethylborane (**21**) with oxygen leads to ethyl radicals (**23**) under considerably milder conditions (even at –78 °C). The resulting initiator radicals react with tributyltin hydride (**24**) or the significantly less toxic tristrimethylsilylsilane (**25**) to give corresponding tin or silicon radicals **31** or **32**, which initiate radical dehalogenation of the substrate alkyl halide **26**. The released carbon radical **29** is now in a position to undergo intramolecular radical cyclization and then abstract a hydrogen atom from an available hydride source, **24** or **25**. This leads again to a heteroatom radical (**31**, **32**), completing the cycle of the chain reaction.[9]

Problem

Explain the observed stereoselectivity in the following radical dehalogenation:

1.0 eq $(Me_3Si)_3SiH$, 0.1 eq Et_3B, C_6H_6, RT; 91%

6 → 7 + C-7-epimer; ds = 13 : 1

Tip

- Draw a three-dimensional representation of the structure of the resulting compound **7** analogous to the structure of compound **6**. Then analyze the steric surroundings of the diatopic chlorine atoms.

Solution

The exceptionally sterically demanding trimethylsilyl radical selectively removes an equatorial chlorine atom. Attack on axial chlorine atoms is severely limited here by the concave molecular geometry. Trialkyl- and triarylstannane were also tried, but they failed to distinguish between the chlorines and instead produced a 1 : 1 epimeric mixture at C-7.

Problem

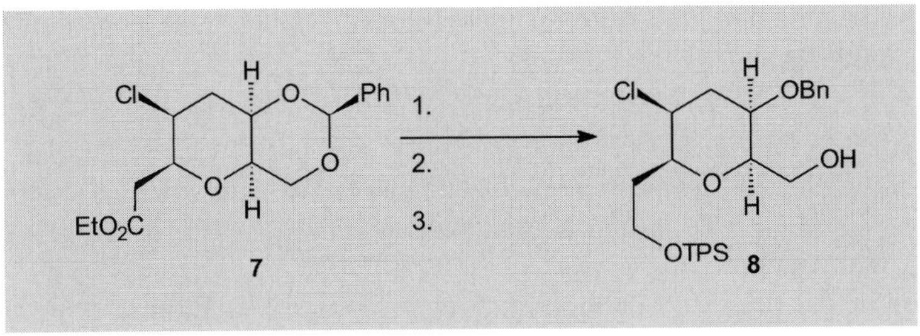

Tips

- The first step accomplishes a prerequisite for the second step, which is formation of a silyl ether.
- Selective cleavage of the acetal is achieved with a reagent similar in type to one already introduced in this chapter.

Solution

The first step is reduction of the ester group to give primary alcohol **33**, which is subsequently protected as the *tert*-butyldiphenylsilyl ether **34**. The benzylidene acetal is selectively cleaved with Lewis-acid activation under reductive conditions to generate protected secondary alcohol **8**.

1. $LiAlH_4$, THF.
2. TPSCl, DMAP, imidazole, 90% over two steps.
3. 1.0 eq. $NaBH_3(CN)$, 1.0 eq. $TiCl_4$, CH_3CN, 0 °C, 93%.

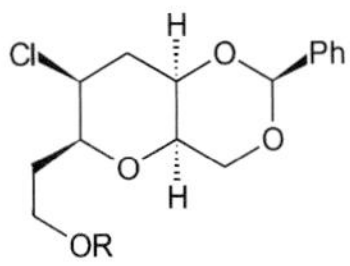

33: R = H
34: R = TPS

Discussion

This reagent combination in a method developed by *Seebach* for selective acetal cleavage is consistent with the presence of ester groups in the molecule.[10] Use of a Lewis acid for activation always leads to transformation of the more highly substituted alcohol into a benzyl ether.

Problem

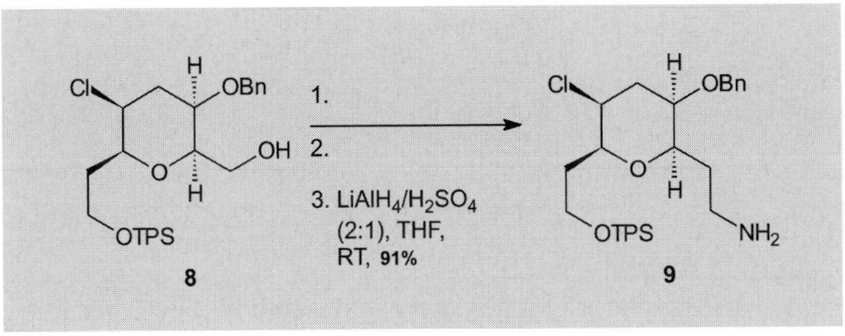

3. $LiAlH_4/H_2SO_4$ (2:1), THF, RT, 91%

Tips

- The number of carbon atoms present is altered in one of the sidechains.
- C_1-Extension of the sidechain and introduction of a nitrogen atom occur simultaneously.
- All this transpires in the course of an S_N2 reaction.
- The first step accomplishes a prerequisite to the nucleophilic substitution.
- In the final step, lithium aluminum hydride and sulfuric acid react to give a Lewis-acidic reducing agent.

Solution

Cl H OBn R O H OTPS

35: R = OTf
36: R = CN

The first step incorporates generation of a triflate leaving group in **35**, which is then replaced by a cyanide ion to give **36**. Lithium aluminum hydride and sulfuric acid react to form aluminum hydride, capable of selectively reducing the nitrile to primary amine **9**.

1. Tf_2O, Py/CH_2Cl_2, 0 °C.
2. KCN, 18-crown-6, 92% over two steps.
3. $LiAlH_4/H_2SO_4$ (2 : 1), THF, 91%.

Discussion

Lithium aluminum hydride as a reducing agent is too basic for use here, since it would cause deprotonation at the position α to the nitrile group. Aluminum hydride reacts extremely slowly with halogen substituents, so the chlorine atom survives this reduction method unscathed.[11]

Problem

Cl H OBn O H OTPS NH_2 **9**

1. 1.5 eq ^{t}BuONO, 2.0 eq $CuBr_2$, CH_3CN, 65 °C, 64%
2. BCl_3, CH_2Cl_2, 93%
3. 1.2 eq **17**, 1.5 eq NMM, CH_2Cl_2, RT, 95%

10

Tips

- The first step produces the byproducts N_2, ^{t}BuOH, H_2O, and CuBr (2.0 eq.).
- A selective ether cleavage occurs in the second step.
- An example of the third transformation has already been encountered in this chapter.

Solution

The first reaction is an oxidative deamination.[12] Proper choice of the copper(II) salt makes it possible to construct a geminal dihalide. The stoichiometric equation for this reaction is:

$$RCH_2NH_2 + {}^tBuONO + 2\,CuBr_2 \xrightarrow[-2\,CuBr]{-N_2,\ -H_2O} {}^tBuOH + RCHBr_2$$

37: R = Bn
38: R = H
10: R = CH=CHCO$_2$Et

Boron trichloride acting as a Lewis acid cleaves benzylic ether **37** selectively to **38**.[13] The third reaction, like the transformation of **4** to **5**, leads to a *trans*-β-alkoxyacrylate.[7]

Problem

10 → **11** (1.2 eq ^{n}Bu$_3$SnH, 0.2 eq AIBN, C$_6$H$_6$, reflux; 75%)

Tips

- A tin radical abstracts a bromine atom, leaving a carbon radical. The remaining bromine atom is oriented preferentially toward the convex side of the molecule.
- Reaction proceeds through a chair-like transition state.
- The fragment O-C-13-acrylate ester assumes an *s-trans* conformation.

Solution

Carbon radical **39** reacts highly selectively from a chair-like transition state to give isomerically pure product **11**. The 6-*exo-trig* cyclization proceeds from a reactive conformation in which steric interactions in the course of ring closure are minimized.

s-trans
convex side
39

Problem

Cl, H, O, CO₂Et, Br, H, O, OTPS, 11 → 1. 2. 3. → Cl, H, O, Br, H, O, OTPS, 12

Tips

- The first step leads to a primary alcohol. Which of the following reagents would you choose: a) $NaBH_4$; b) B_2H_6; or c) $LiAlH_4$?
- The primary alcohol is halogenated in an *Appel* reaction.[14]
- A powerful H-nucleophilic reagent is employed in the third step.

Solution

Lithium aluminum hydride as reducing agent permits formation of alcohol **40** in the first step. This is then converted to iodide **41**. Another reduction with a nucleophilic boron reagent produces an ethyl group at C-13 and thus structure **12**.

1. $LiAlH_4$, THF.
2. I_2, imidazole, PPh_3.
3. $LiEt_3BH$, 89% over three steps.

40: R = OH
41: R = I

Problem

12 → 1. 2. → 13

Tips

- The first step accomplishes cleavage of the silyl ether.
- A mild oxidation process generates compound **13**.

Solution

Here the use of a highly unusual reagent leads to successful cleavage of the *tert*-butyldiphenylsilyl ether. Alcohol **42** is prepared with *p*-TsOH. In the second step an oxidation results in aldehyde **13**.[15]
1. *p*TsOH.
2. $SO_3 \cdot$pyridine, DMSO, NEt_3, CH_2Cl_2, 91% over two steps.

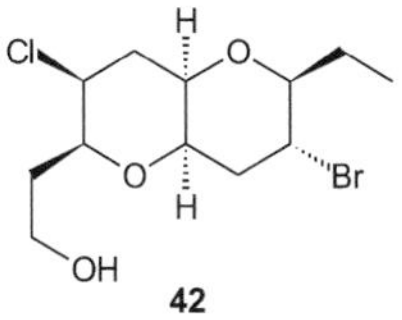

42

Discussion

Sulfur trioxide is one of many reagents [e.g., DCC, Ac_2O, $(COCl)_2$, TFAA] with which DMSO can be activated as an oxidizing reagent for alcohols. Oxalyl chloride has found the widest application in the reaction named after *Swern*.[16] Structure **43** represents the activated species in the above oxidation.

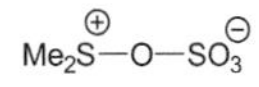

$Me_2\overset{\oplus}{S}{-}O{-}SO_3^{\ominus}$

43

Problem

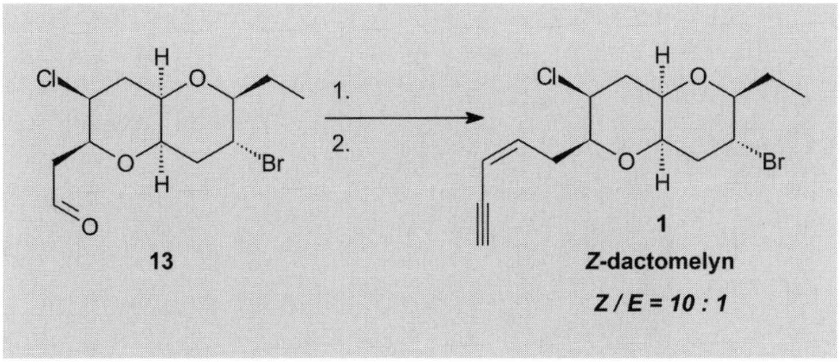

13 → 1., 2. → **1** **Z-dactomelyn** ***Z/E* = 10 : 1**

Tips

- The first transformation introduces all the still missing carbon atoms.
- The process is a nucleophilic addition.
- The double bond is formed by an olefination reaction involving a silicon atom.
- An alkyne lithiated in the propargylic position bears two silyl groups.
- The second step is a protecting-group operation induced by fluoride ion.

Solution

The reagent employed in the first reaction, with structure **44**, is from a method developed by *Corey* for the synthesis of conjugated enynes.[17] Addition of TBAF leads to quantitative formation of target structure **1**.
1. 44, THF, –78 °C, 68%.
2. 1.5 eq. TBAF, THF, RT, 99%.

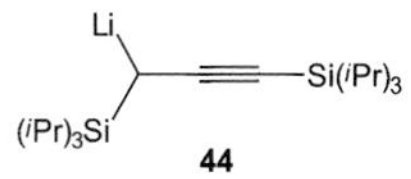

44

Discussion

Preferred formation of a double bond with the *Z* configuration is explained by Corey as follows. In solution there exists an equilibrium between lithiated alkyne **44** and allene **45**. The latter adds to the aldehyde with minimization of steric interactions to give intermediate **46**. Two stereocenters at C-3 and C-4 are thereby created in a single step. Simple diastereoselectivity is observed, leading to preferential formation of diastereomer **46a** over **46b** because of steric interactions.

These isomers differ only in the relative configurations of the two stereocenters. The transition state leading to **46a** is more favorable because here there are no interactions between the bicyclic system and the alkyne.

After rotation about the newly formed bond, nothing stands in the way of selective *syn*-elimination from **47** – a possible intermediate of a base-induced *Peterson* elimination – to *Z*-olefin **48**.[18]

14.4 Summary

In the synthesis presented here of *Z*-dactomelyn (**1**), three key radical steps permitted stereoselective construction of a bispyran skeleton with the development of halogen-substituted stereogenic centers. Starting from (–)-diethyl tartrate (**2**), the target compound was prepared in 26 steps with an overall yield of 0.04%.

14.5 References

1 Y. Gopichand, F.J. Schmitz, J. Shelly, A. Rahman, D. van der Helm, *J. Org. Chem.* **1981**, *46*, 5192.
2 J.G. Hall, J.A. Reiss, *Aust. J. Chem.* **1986**, *39*, 1401.
3 A.P. Kozikowski, J. Lee, *J. Org. Chem.* **1990**, *55*, 863.
4 E. Lee, C.M. Park, J.S. Yun, *J. Am. Chem. Soc.* **1995**, *117*, 8017.
5 A.H. Al-Hakim, A.H. Haines, C. Morley, *Synthesis* **1985**, 207.
6 H. Taguchi, H. Yamamoto, H. Nozaki, *J. Am. Chem. Soc.* **1974**, *96*, 3010.
7 E. Winterfeldt, H. Preuss, *Chem. Ber.* **1966**, *99*, 450; J.B. Hendrickson, R. Rees, J.F. Templeton, *J. Am. Chem. Soc.* **1964**, *86*, 107; E. Lee, J.S. Tae, C. Lee, C.M. Park, *Tetrahedron Lett.* **1993**, *34*, 4831.
8 A.L.J. Beckwith, C.H. Schiesser, *Tetrahedron* **1985**, *41*, 3925; A.L. J. Beckwith, C.J. Easton, A.K. Serelis, *J. Chem. Soc., Chem. Commun.* **1980**, 482; A.L.J. Beckwith, T. Lawrence, A.K. Serelis, *J. Chem. Soc., Chem. Commun.* **1980**, 484.
9 U. Koert, *Angew. Chem. Int. Ed. Engl.* **1996**, *35*, 405; *Angew. Chem. Int. Ed. Engl.* **1996**, *35*, 405; B. Giese, *Angew. Chem. Int. Ed. Engl.* **1989**, *28*, 969; *Angew. Chem. Int. Ed. Engl.* **1989**, *28*, 969; C.P. Jasperse, D.P. Curran, T.L. Fevig, *Chem. Rev.* **1991**, *91*, 1237; B. Giese, B. Kopping, *Tetrahedron Lett.* **1989**, *30*, 681; C. Chatgilialoglu, B. Giese, B. Kopping, *Tetrahedron Lett.* **1990**, *31*, 6013; H.C. Brown, M.M. Midland, *Angew. Chem.* **1972**, *84*, 702; *Angew. Chem. Int. Ed. Engl.* **1972**, *11*, 692; K. Tamao, K. Nagata, Y. Ito, K. Maeda, M. Shiro, *Synlett* **1994**, 257; K. Nozaki, K. Oshima, K. Utimoto, *J. Am. Chem. Soc.* **1987**, *109*, 2547; A.G. Myers, D.Y. Gin, D.H. Rogers, *J. Am. Chem. Soc.* **1993**, *115*, 2036; T.B. Lowinger, L. Weiler, *J. Org. Chem.* **1992**, *57*, 6099.
10 G. Adam, D. Seebach, *Synthesis* **1988**, 373.
11 N.M. Yoon, H.C. Brown, *J. Am. Chem. Soc.* **1968**, *90*, 2927.
12 M. Doyle, B. Siegfried, *J. Chem. Soc., Chem. Commun.* **1976**, 433.
13 D.R. Williams, D.L. Brown, J.W. Benbow, *J. Am. Chem. Soc.* **1989**, *111*, 1923.
14 R. Appel, *Angew. Chem. Int. Ed. Engl.* **1975**, *14*, 801; *Angew. Chem. Int. Ed. Engl.* **1975**, *14*, 801.
15 J.R. Parikh, W. von E. Doering, *J. Am. Chem. Soc.* **1967**, *89*, 5505.
16 A.K. Sharma, D. Swern, *Tetrahedron Lett.* **1974**, 1503, K.A. Sharma, T. Ku, A.D. Dawson, D. Swern, *J. Org. Chem.* **1975**, *40*, 2758.
17 E.J. Corey, C. Rücker, *Tetrahedron Lett.* **1982**, *23*, 719.
18 D.J. Peterson, *J. Org. Chem.* **1968**, *33*, 780; review: D.J. Ager, *Org. React.* **1990**, *38*, 1.

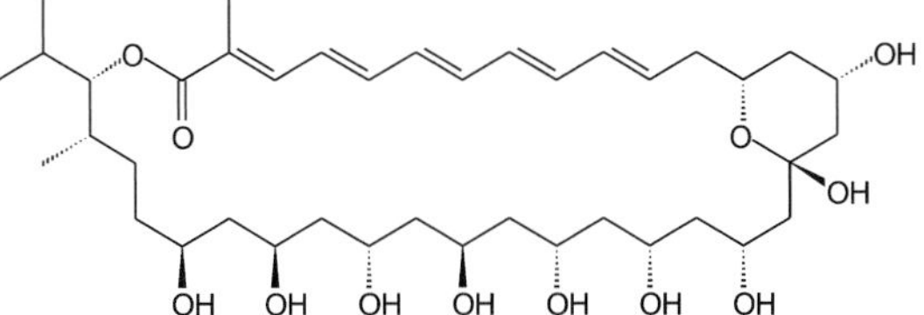

15

Maehr's Roflamycoin: Rychnovsky (1994)

15.1 Introduction

The over 200 known members of the class of compounds known as macrolide antibiotics[1] includes several active substances, such as erythromycin (see Chapter 6), that are highly prized as backup medications: drugs to which one can resort when resistance develops to more conventional pharmaceuticals.[2]

The macrolide antibiotic roflamycoin (**2**) was isolated from *Streptomyces roseoflavus* and was initially given the name flavomycoin.[3] The constitution of this natural product was established in 1981 by *Schlegel*.[4] Its structure is characterized by the presence of two chains, one a polyene and the other a polyol. The compound is also a hemiacetal.

1: R^1 = H, R^2 = OH, R^3 = OH, R^4 = H, R^5 = H, R^6 = OH ***Maehrs* roflamycoin**
2: R^1 = OH, R^2 = H, R^3 = H, R^4 = OH, R^5 = OH, R^6 = H **roflamycoin**

Maehr was successful in 1989 in determining by X-ray crystallography the structure of another polyene carbonyl compound, roxaticin. On the basis of similarities between this substance and roflamycoin he suggested for the latter's polyol chain the relative and absolute configurations shown in **1**. For practical reasons, *Rychnovsky* selected as a synthetic target the compound with the configurations at C-13 and C-15 illustrated above. A comparison of spectroscopic data from the synthetic product, here referred to as *Maehr*'s roflamycoin, with that of the substance isolated from nature indicated that the two differed in their configurations at C-21, C-25, and C-27. The synthesis of *Maehr*'s roflamycoin (**1**), described in this chapter, served to effectively rule out *Maehr*'s proposed stereochemistry for the natural product roflamycoin (**2**).

15.2 Overview

3

4

$AlCl_3$, nitrobenzene,
1,2-dichloroethane;
0 °C bis 60 °C

5

1.
2.

6

1. $^nBu_3Sn$ OBn 7
^nBuLi, BF_3*OEt_2 (kat.)
2. ^nBuLi, 8

59%

9

1.
2.
3.
4.
5.

OTBS
NC
10
OBn
CH_3

44%

lithium pyrrolidide, THF, -78 °C,
2.2 eq **11**, DMPU, -78 °C→ -20 °C

11

12

35 OBn
CN
31
CN
27
CN
21
OTBS
OBn
19
CN
13
11
CH_3

13

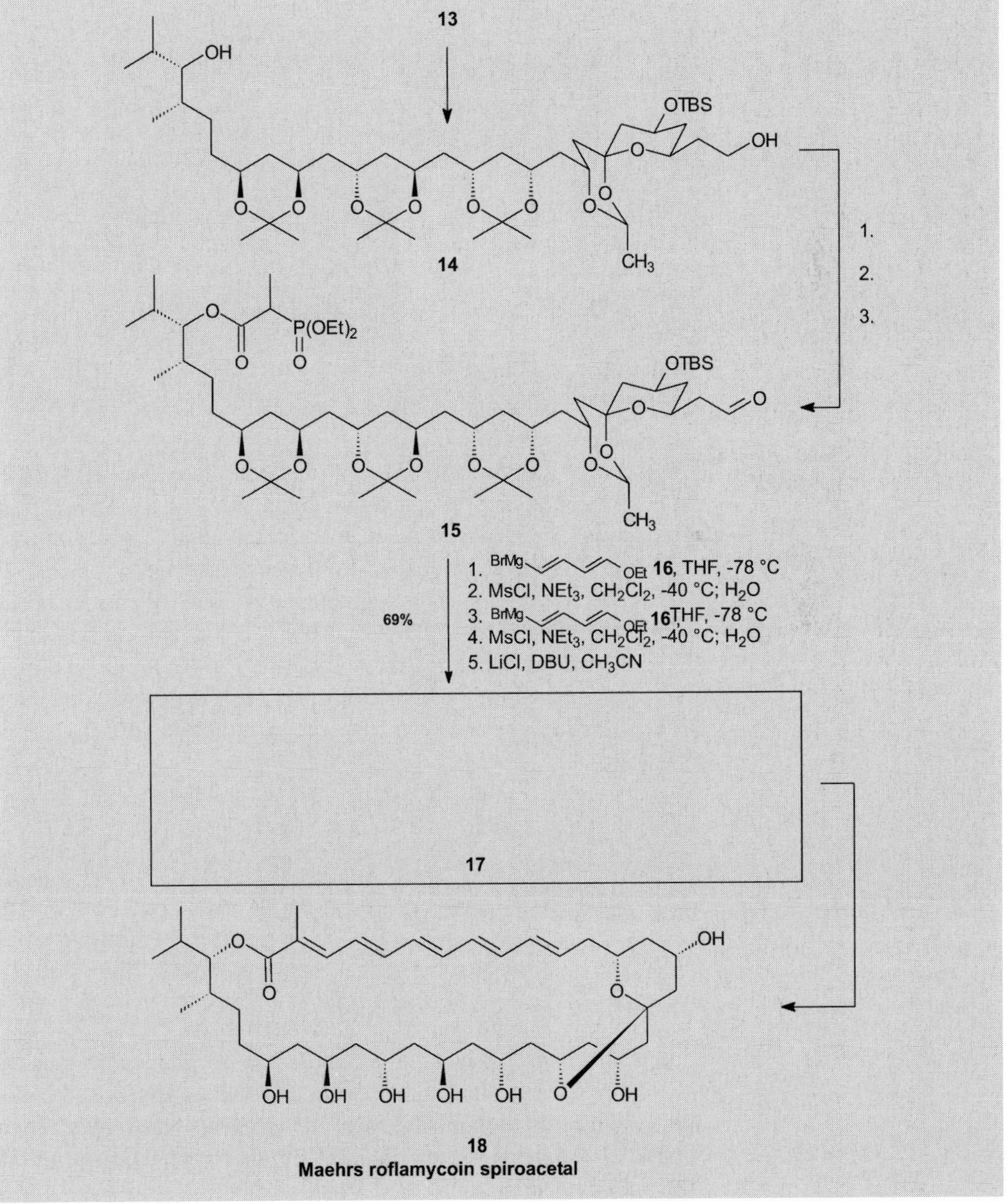
13
OH
OTBS
OH
O
O O
O O
O O
O
O
14
CH3
1.
2.
3.
O
P(OEt)2
O
O
OTBS
O
15
CH3
1. BrMg OEt 16, THF, -78 °C
2. MsCl, NEt3, CH2Cl2, -40 °C; H2O
3. BrMg OEt 16, THF, -78 °C
4. MsCl, NEt3, CH2Cl2, -40 °C; H2O
5. LiCl, DBU, CH3CN
69%
17
OH
OH OH OH OH OH O OH
18
Maehrs roflamycoin spiroacetal

15.3 Synthesis

Problem

3 → 4, $AlCl_3$, nitrobenzene, 1,2-dichloroethane, 0 °C → 60 °C → 5

Tips

- In the presence of the Lewis acid aluminum trichloride, acetylacetone (**3**) carries out a nucleophilic attack on chloroacetyl chloride (**4**).
- C-3 is the most acidic position in acetylacetone (**3**).
- This is an equilibrium reaction in which a molecule of acetyl chloride is eliminated.
- Product **5** contains two chlorine atoms.

Solution

A reaction of acetylacetone (**3**) with chloroacetyl chloride (**4**) takes place in the presence of aluminum trichloride. In this process, an acetyl group is exchanged for a chloroacetyl group. Only atom C-3 in product **5** is therefore derived from acetylacetone (**3**).

A reaction mechanism is conceivable in which the intermediate adduct **19** arises from chloroacetyl chloride (**4**) and acetylacetone (**3**). This intermediate could then eliminate acetyl chloride via **20** to give the monohalogenated intermediate **21**. Further reaction to **5** might subsequently occur with a second molecule of acetyl chloride (**4**).

19 ⇌ 20 ⇌ (- AcCl) 21

22

Separation of the products is accomplished by shaking an ethereal solution of the reaction mixture with saturated copper acetate solution, which leads to precipitation of the ether-insoluble copper complex **22**. Diketone **5** is then obtained by treating this complex with 10% sulfuric acid.[7]

Problem

Tips

- Both carbonyl groups are hydrogenated enantioselectively in the first step to produce a diol.
- A chiral ruthenium catalyst is used in the hydrogenation.
- A base is introduced in the second step, in which two nucleophilic substitutions take place.

Solution

1,3-Diketone **5** is reduced to diol **23** by the method of *Noyori*:[8] hydrogenation with catalysis by the chiral ruthenium-BINAP complex $\{[(S)\text{-BINAP}]RuCl_2\}_2\text{-}NEt_3$.

α-Halohydroxy compounds like **23** readily cyclize under basic conditions to give the corresponding epoxides, so here the introduction of potassium hydroxide in diethyl ether leads to the C_2-symmetric diepoxide **6**.[7]

1. $\{[(S)\text{-BINAP}]RuCl_2\}_2\text{-}NEt_3$ (cat.), H_2, (86 bar), MeOH, 102 °C, 64%, >97% *ee*.
2. KOH, Et_2O, 89%.

Problem

Tips

- Benzyloxymethyltributylstannane (**7**) undergoes metal exchange with butyllithium. What results is tetrabutylstannane and a lithiated benzyl ether.
- The boron trifluoride complex serves to activate the epoxide.
- Epoxide **6** is attacked nucleophilically by the lithiated benzyl ether.
- A secondary alcohol is obtained in the first step after aqueous workup.

- The second reaction involves a nucleophilic attack on the second epoxide group.
- ^{n}BuLi deprotonates the alcohol obtained in the first reaction and also lithiates the added dithiane **8**.

Solution

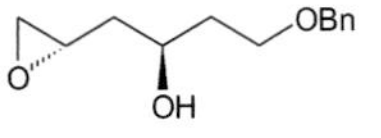

OBn
OH
24

The lithiated benzylic ether prepared from **7**[9] is a carbon nucleophile. Under boron trifluoride catalysis[10] it attacks diepoxide **6**. In the process, the alkyllithium reagent adds in a stepwise fashion to diepoxide **6**, whereby the first addition is significantly more rapid than the second.[7] One thus obtains only the monoadduct **24**.

Nucleophilic attack of lithiated dithiane **8** on the lithium alkoxide of **24** gives 1,3-diol **9**.

Problem

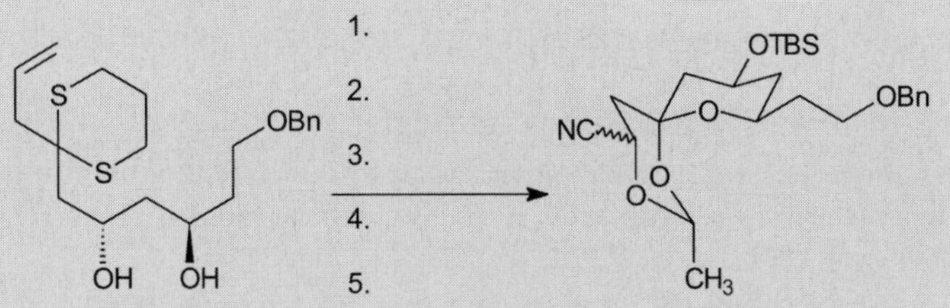

S S OBn OH OH **9** 1. 2. 3. 4. 5. OTBS NC O OBn O O CH_3 **10**

Tips

OH
O
OBn
OMe
25α

- The thioacetal structure is first removed oxidatively.
- Deprotection of the keto function is carried out in methanol. The reaction product then forms a 6-membered cyclic acetal.
- The product of the first reaction is present in the form of acetal **25α**.
- The second transformation is removal of the silyl protecting group.
- The double bond is oxidized in the third step, causing the molecule to be shortened by one carbon atom.
- In the fourth reaction the aldehyde function created in the previous step is converted into a cyanohydrin.
- In the fifth step a spiroacetal structure is constructed. This occurs under acid catalysis with addition of acetaldehyde.

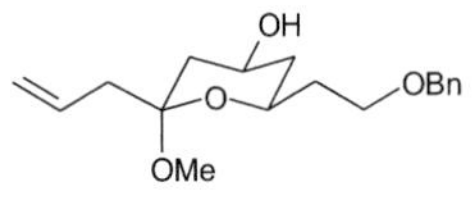

Solution

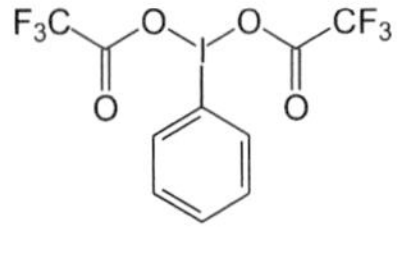

F_3C O I O CF_3 O O
26

The first reaction step involves a method developed by *Stork*: use of the hypervalent-iodine species bis(trifluoroacetoxy)iodobenzene (**26**), which effects oxidative removal of the dithiane.[11] Methylacetal **25** α is formed in methanol solution in the presence of traces of acid. Subsequent silylation of the secondary alcohol is accomplished using TBS-triflate with lutidine as base. The third reaction is an ozonolysis. After reductive workup with dimethyl sulfide the aldehyde **27** is obtained, which in the fourth step is transformed into the corresponding cyanohydrin **28** (or the silyl ether thereof) with trimethylsilyl cyanide under catalysis by the KCN/18-crown-6 complex.[12] With acetaldehyde as reagent and catalytic amounts of CSA the spiroacetal **10** forms, presumably via intermediate **29**.

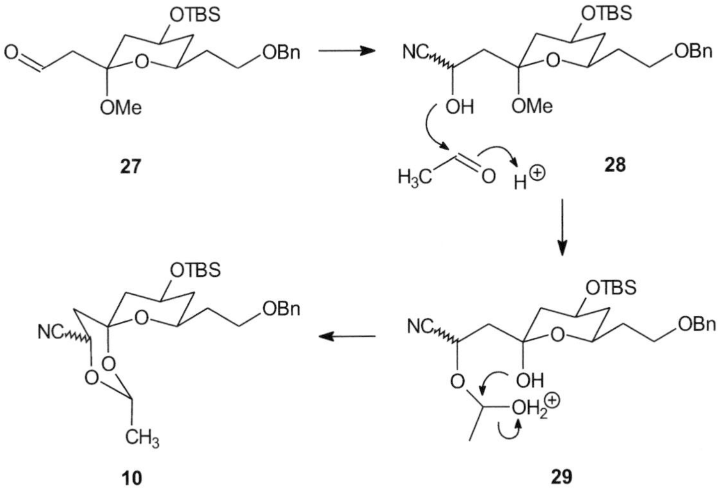

1. Bis(trifluoroacetoxy)iodobenzene (**26**), MeOH.
2. TBSOTf, lutidine, 81% over two steps.
3. O_3, SMe_2.
4. TMSCN, KCN/18-crown-6.
4. Acetaldehyde, CSA, 42% over three steps.

Discussion

Compared with *O,O*-acetals, thioacetals are much more stable to acid as a result of the lower Brønsted basicity of sulfur. To increase the leaving-group capacity of sulfur, and thus to simplify the cleavage of thioacetals, three different methods can in principle be invoked:[13]

1. One can use metal coordination to take advantage of the high affinity of heavy-metal cations for sulfur. A salt of Hg(II), Ag(I), Ag(II), Cu(II) or Tl(III) would thus be introduced.
2. In some procedures, thioacetal cleavage occurs after alkylation at the sulfur atom with a reactive alkylating reagent such as methyl iodide, trialkyloxonium tetrafluoroborate (*Meerwein* salt), or ethyl trifluoromethanesulfonate to give a trialkylsulfonium salt.
3. The third method is oxidative deprotection, which initially results in oxidation to a sulfoxide. Useful oxidizing agents include

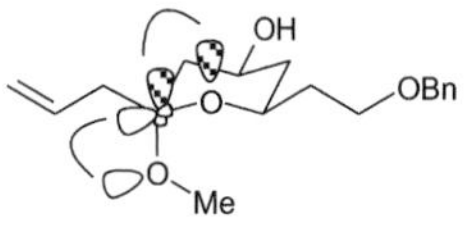

25α

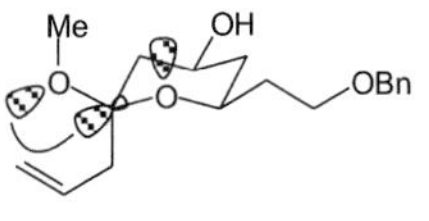

25β

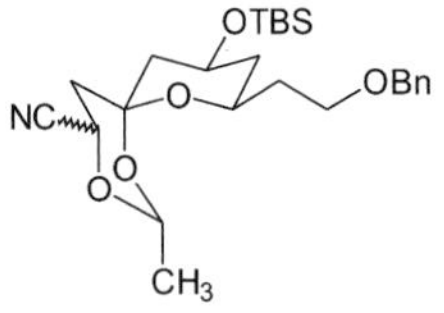

10

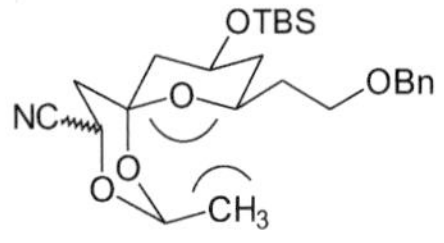

10a

the elemental halogens as well as NCS and NBS, N-chlorobenzotriazole, chloramine-T, MCPBA, periodic acid, and such heavy-metal oxidants as thallium(III) nitrate or CAN.

The advantage of the procedure introduced by *Stork* with bis(trifluoroacetoxy)iodobenzene (**26**) derives from the associated mild reaction conditions. Esters, thioesters, nitriles, secondary amides, amines, alcohols, halides, alkenes, and alkynes are all compatible with this transformation.

Acetal **25** is present as an α-anomer: i.e., with an axial methoxy group. In explaining this stereochemical phenomenon, the so-called anomeric effect, it is common to rely on an orbital-interaction argument.[14] Thus, overlap of the nonbonding, occupied n-orbitals of the acetal oxygen atoms with unoccupied σ* orbitals from the other C-O bond is accompanied by delocalization of electrons and thus stabilization. In the α-configuration with axial oxygen, two such interactions are possible, whereas an equatorially oriented oxygen can engage in only a single (n-σ*) overlap, so the latter is the energetically less favorable structure. An analogous argument applies to spiroacetal **10**.

The methyl group derived from acetaldehyde in **10** occupies an equatorial position. Steric arguments can be used to explain this result, supported by the fact that in a similar model system with acetone no spiroacetalization occurs.[15]

In the structures established in this way for compounds **10**, **25**, **27**, **28**, and **29** the stereochemistry at C-17 is determined by the configuration at C-13.

Problem

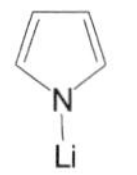

lithium pyrrolidide

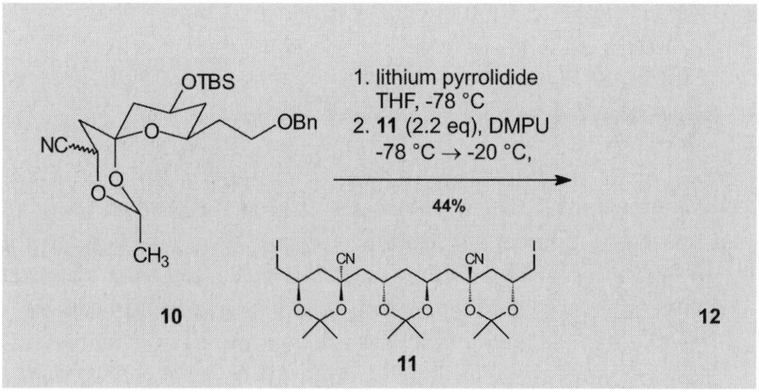

1. lithium pyrrolidide THF, -78 °C
2. **11** (2.2 eq), DMPU -78 °C → -20 °C,

44%

10 **11** **12**

Tips

- Spiroacetal **10** is first deprotonated.
- The proton α to the nitrile group is abstracted with lithium pyrrolidide.
- After addition of the C_2-symmetric polyol building block **11**, C-C coupling is achieved through a nucleophilic substitution reaction.

Solution

The configuration of the α-lithiated nitrile is not stable at C-19, as is easily seen from resonance structure **30**. Thus, the unselectively created stereogenic center at C-19 in spiroacetal **10** effectively becomes a prochiral center. Alkyl iodide **11**, introduced in considerable excess to prevent double alkylation, approaches from the sterically less hindered side, as a result of which the nitrile group in product **12** assumes an axial position.[16] In this way the configuration at C-19 is ultimately determined also by that at C-13.

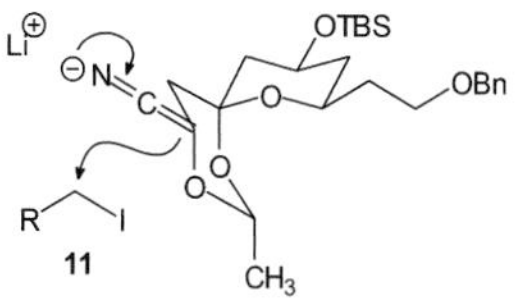

Discussion

One of the methods frequently employed in this natural product synthesis is application of "umpoled" reagents and substrates. Coupling to **12** has been prearranged, in that the electrophilic aldehyde was changed into a nucleophile by conversion to metallated cyanohydrinacetal **30**.[17] The reactivity of the carbonyl group is thus reversed. Another "umpoled" reagent has already been utilized in this synthesis in the form of the lithiated dithane **9**.[18]

Problem

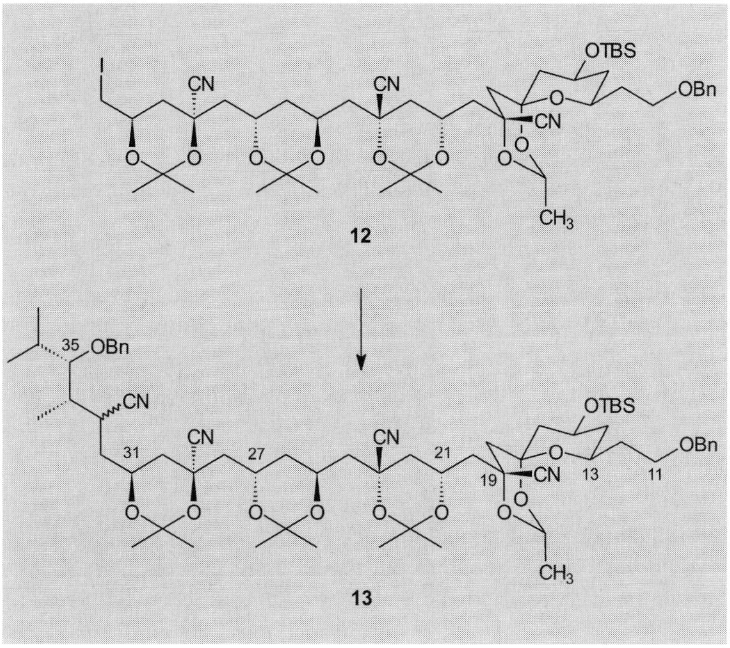

Tips

- An extension of the carbon skeleton is achieved through an approach very similar to the preceding reaction.
- The reagent utilized is a lithiated nitrile.

Solution

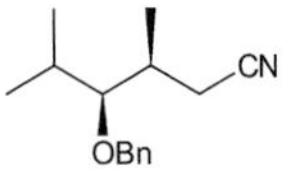

Enantiomerically pure nitrile **31** is deprotonated with lithium diethylamide in THF. The subsequently added iodide **12** is then alkylated in the presence of DMPU.
31 (3.6 eq.), $LiNEt_2$, THF, –78 °C; **12**, DMPU, –78 °C → –10 °C, 95%.

Problem

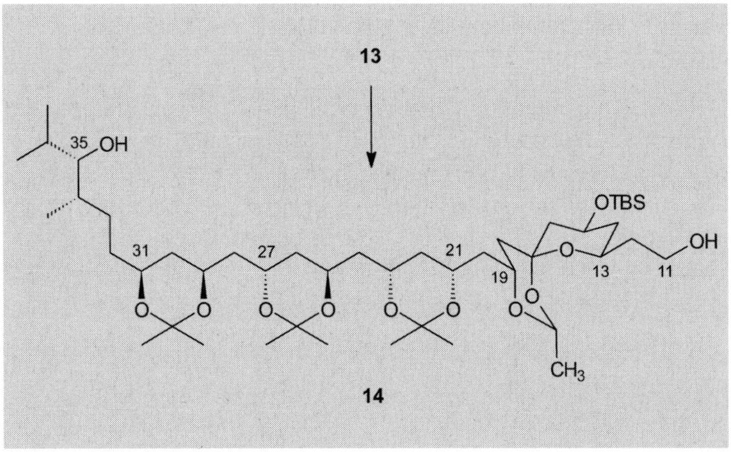

Tips

- The benzyl and cyanide groups are removed in a single step.
- The process is reductive in nature.
- The reducing agent is a metal in the liquid phase.

Solution

Under *Birch* reduction conditions not only are the benzyl groups at C-11 and C-35 cleaved, the nitrile groups at C-19, C-23, C-29, and C-33 also undergoes replacement by hydrogen in a reductive decyanation. In this case the procedure involves using lithium in liquid ammonia with THF as a cosolvent (see Chapter 12).
Li, NH_3 (l), THF, 49%.

Discussion

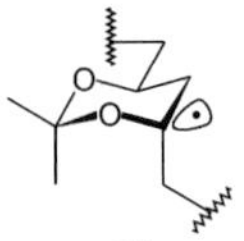

Subsequent to decyanation the newly introduced hydrogen atoms assume the axial positions of the replaced nitrile groups. *Rychnovsky* examined the high stereoselectivity of this reaction both experimentally and theoretically.[16,19] Decyanation begins with the elimination of a nitrile radical anion, which is reduced to a cyanide anion. The remaining alkyl radical is further reduced and after protonation gives an alkane. Results of the study argue for intermediate appearance of the pyramidal, quasi-axial radical **32b**, which is in rapid equilibrium with the quasi-equatorial form **32a**. Incorporation of an electron leads to a configurationally stable axial anion that is then protonated with retention of configuration.

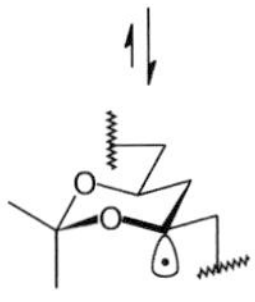

Problem

Tips

- In the first step both hydroxy groups in compound **14** are esterified.
- Esterification is accomplished with diethylphosphonopropionic acid **33**.
- The ester of the primary hydroxy group at C-11 is selectively hydrolyzed in the second step.
- The third reaction is oxidation of the primary hydroxy group.
- A hypervalent iodine species is used for the oxidation.

Solution

With the aid of BOP [benzotriazol-1-yl-oxytris(dimethylamino) phosphonium hexafluorophosphate, *Castro*'s reagent][20] in the presence of DMAP it is possible to transform the free carboxylic acid **33** into activated intermediate **34** in preparation for subsequent esterification with **14**. Other methods for esterification include application of DCC and the procedures introduced by *Yamaguchi* and *Mukaijama* (see Chapters 5, 6, and 9).

The reaction product obtained in this way is treated with methanol saturated with ammonia in order selectively to cleave the ester of the primary hydroxy group at C-11. Subsequent oxidation to aldehyde **15** is accomplished with the *Dess-Martin* periodinane.

(EtO)2P OH 33 BOP, DMAP (EtO)2P NMe2 NMe2 PF6 NMe2 34 HO—R 14

1. **33**, BOP, DMAP, CH_2Cl_2.
2. NH_3, MeOH.
3. *Dess-Martin* periodinane, 74% over three steps.

Problem

P(OEt)2 OTBS CH3 15

1. BrMg OEt, THF, -78 °C → 0 °C **16**
2. MsCl, NEt_3, CH_2Cl_2, -45 °C; H_2O
3. BrMg OEt, THF, -78 °C → 0 °C **16**
4. MsCl, NEt_3, CH_2Cl_2, -45 °C; H_2O
5. LiCl, DBU, CH_3CN

69%

17

Tips

- In the first reaction, compound **15** is coupled with the C_4 building block **16**.
- Treatment of aldehyde **15** with *Grignard* reagent **16** produces a secondary alcohol.
- In the second step the enol ether introduced with *Grignard* reagent **16** is converted into an aldehyde function. For this purpose the secondary alcohol prepared in the first step is mesylated and thus transformed into a good leaving group.

- The enol ether is then hydrolyzed, taking advantage of its conjugation.
- The first two steps are repeated in the third and fourth steps.
- The fifth step accomplishes a cyclization using a variant of the *Horner-Wadsworth-Emmons* reaction.

Solution

Relying on the *Wollenberg* method,[21] aldehyde **15** is converted with *Grignard* reagent **16** in the first step into the α,β- and γ,δ-unsaturated hydroxy compound **35**, which has been extended by a C_4 unit. After mesylation of the resulting secondary alcohol to **36**, the enol ether is hydrolyzed via **37** to aldehyde **38**. Repeating this procedure leads to the C_8-extended aldehyde **39**.

R O 15 + BrMg OEt 16 → OH R OEt 35 ↓ MsCl, NEt_3 → OMs R OEt 36 → H_2O → R O–H ⊕OEt H 37 ↓ R O 38

Since the resulting polyene structure is sensitive to strong base, macrocyclization is carried out with a variant on the *Horner-Wadsworth-Emmons* reaction involving lithium chloride and DBU.

Lithium chloride makes it possible to increase the acidity of phosphonate **39** considerably, so that deprotonation and a subsequent *Horner-Wadsworth-Emmons* reaction[22] (see Chapters 5 and 8) is possible even with the rather mild base DBU. *Roush* and *Masamune*[23] attribute this result to the formation of tight ion pair **40** between the lithium cation and the phosphonate anion. Macrolactonization alone occurs with a yield of 89%.

Base H P(OEt)$_2$ O Li O 39 ↓ P(OEt)$_2$ O Li O 40

Problem

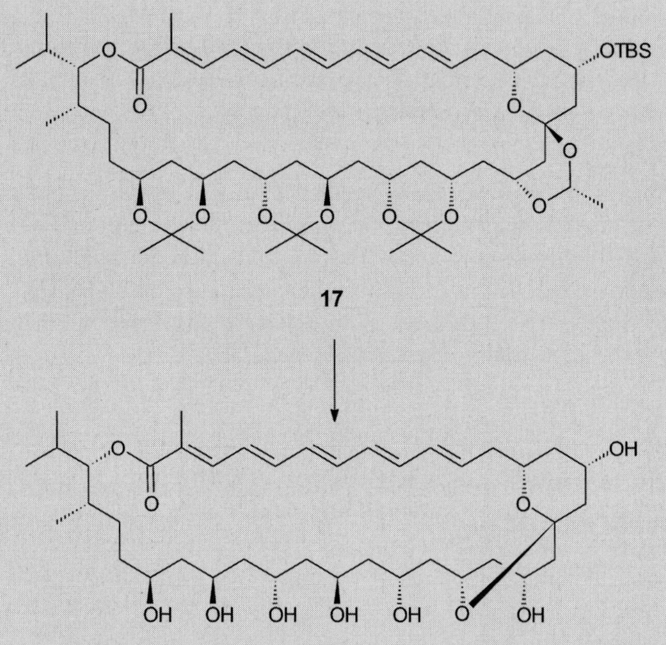

17

18
***Maehrs* roflamycoin spiroacetal**

Tips

- Acetals can be hydrolyzed with acid.
- Silyl ethers are likewise subject to acid hydrolysis.
- The proton source employed is in the form of a solid.

Solution

A reaction with acidic ion-exchange resin Dowex 50W-X1 cleaves both the acetonide protecting group and the silyl ether. Simultaneously there is formed a 6-membered-ring spiroacetal linking C-17 with C-21, thereby producing structure **18**.
Dowex 50W-X1 (H^+), MeOH, 60%.

Discussion

A similar spiroacetal **41** was obtained after treatment of the natural product with an acidic ion-exchange resin.

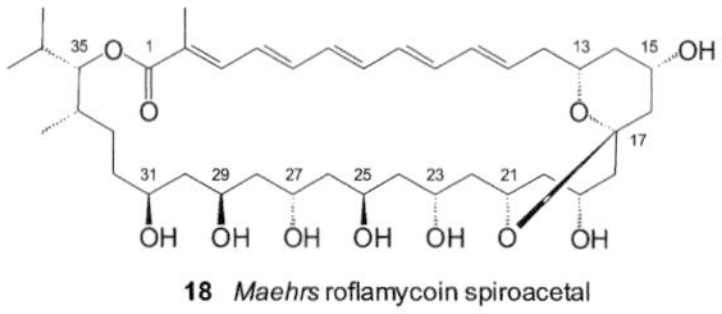

18 *Maehrs* roflamycoin spiroacetal

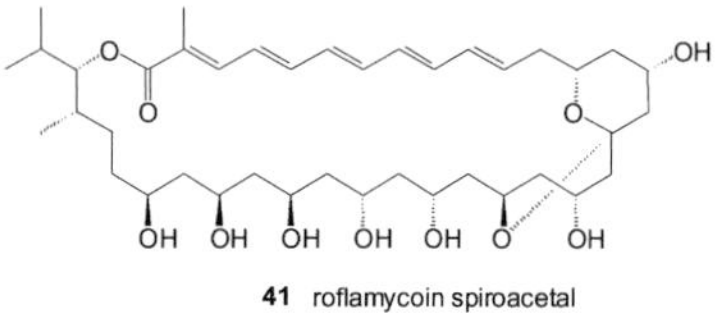

41 roflamycoin spiroacetal

Through a comparison of spectroscopic data from the natural product with that from the synthetic material, *Rychnovsky* ascertained that a stereoisomer of the active substance had been synthesized. Further NMR spectroscopic experiments with various derivatives of the two substances made it possible to establish the actual stereochemistry of roflamycoin. *Rychnovsky*'s research group carried out a synthesis of roflamycoin itself in 1997[24] using a strategy similar to the one described here.

15.4 Summary

Starting from the enantiomerically pure and readily accessible diepoxypentane **4** it proved possible to prepare acetal **15** in 18 synthetic steps.

A central element in the synthetic strategy presented here for the preparation of the chiral polyol chain is the use of chiral cyanohydrin acetals, which are reacted diastereoselectively under substrate control with C_2-symmetric electrophiles.

The *Wollenberg* method was used to construct the sensitive *trans*-polyene structure of roflamycoin.

A variation on the *Horner-Wadsworth-Emmons* reaction was selected for the macrocyclization, since the possibility for macrolactonization – an obvious approach for a lactone – is ruled out due to steric hindrance from the isopropyl group in the position (C-35) adjacent to the lactone functionality.

After removal of the acetonide protecting group the free polyol is obtained as spiroacetal derivative **18**.

15.5 References

1 S.D. Rychnovsky, *Chem. Rev.* **1995**, *95*, 2021; R. Norcross, I. Paterson, *Chem. Rev.* **1995**, *95*, 2041.

2 E. Mutschler, *Arzneimittelwirkungen*, Wissenschaftliche Verlagsgesellschaft Stuttgart 1996, p. 682.

3 R. Schlegel, H. Thrum, *Experientia* **1968**, *24*, 11.

4 R. Schlegel, H. Thrum, *J. Antibiot.* **1971**, *24*, 360; R. Schlegel, H. Thrum, *J. Antibiot.* **1971**, *24*, 368; R. Schlegel, H. Thrum, J. Zielinski, E. Borowski, *J. Antibiot.* **1981**, *34*, 122.

5 H. Maehr, R. Yang, L.-N. Hong, C.-M. Liu, M.H. Hatada, L.J. Todaro, *J. Org. Chem.* **1989**, *54*, 3816.

6 K. Matsui, M. Motoi, T. Nojiri, *Bull. Chem. Soc. Jpn.* **1973**, *46*, 562.

7 S. D. Rychnovsky, G. Griesgraber, S. Zeller, D.J. Skalitzky, *J. Org. Chem.* **1991**, *56*, 5161.

8 R. Noyori, T. Ohkuma, M. Kitamura, H. Takaya, N. Sayo, H. Kumobayashi, S. Akutagawa, *J. Am. Chem. Soc.* **1987**, *109*, 5856; M. Kitamura, T. Ohkuma, S. Inoue, N. Sayo, H. Kumobayashi, S. Akutagawa, T. Ohta, H. Takaya, R. Noyori, *J. Am. Chem. Soc.* **1988**, *110*, 629.

9 W.C. Still, *J. Am. Chem. Soc.* **1978**, *100*, 1481.

10 W. J. Eis, J.E. Wrobel, B. Ganem, *J. Am. Chem. Soc.* **1984**, *106*, 3693.

11 G. Stork, K. Zhao, *Tetrahedron Lett.* **1989**, *30*, 287.

12 D.A. Evans, L.K. Truesdale, G.L. Carroll, *J. Chem. Soc., Chem. Commun.* **1973**, 55.

13 B.-T. Gröbel, D. Seebach, *Synthesis*, **1977**, 357; P.J. Kocienski, *Protecting Groups*, Georg Thieme Verlag, Stuttgart 1994, p. 171.

14 H. Yuasa, Y. Kamata, H. Hashimoto, *Angew. Chem. Int. Ed. Engl.* **1997**, *36*, 868; *Angew. Chem. Int. Ed. Engl.* **1997**, *36*, 868.

15 S.D. Rychnovsky, G. Griesgraber, *J. Chem. Soc., Chem. Commun.* **1993**, 291.

16 S.D. Rychnovsky, S. Zeller, D.J. Skalitzky, G. Griesgraber, *J. Org. Chem.* **1990**, *55*, 5550; S.D. Rychnovsky, G. Griesgraber, *J. Org. Chem.* **1992**, *57*, 1559.

17 G. Stork, S. Maldonado, *J. Am. Chem. Soc.* **1971**, *93*, 5286; J.D. Albright, *Tetrahedron* **1983**, *39*, 3207.

18 S. Hünig, *Chimia* **1982**, *36*, 1.

19 S.D. Rychnovsky, J.P. Powers, T.J. LePage, *J. Am. Chem. Soc.* **1992**, *114*, 8375.

20 Y. Chapleur, B. Castro, *J. Chem. Soc., Perkin Trans. 1* **1980**, 2683.

21 R.H. Wollenberg, *Tetrahedron Lett.* **1978**, *19*, 717.

22 R. Brückner, *Reaktionsmechanismen*, Spektrum Akademischer Verlag, Heidelberg 1996, p. 320.

23 M.A. Blanchette, W. Choy, J.T. Davis, A.P. Essenfeld, S. Masamune, W.R. Roush, T. Sakai, *Tetrahedron Lett.* **1984**, *25*, 2183.

24 S.D. Rychnovsky, U.R. Khire, G. Yang, *J. Am. Chem. Soc.* **1997**, *119*, 2058.

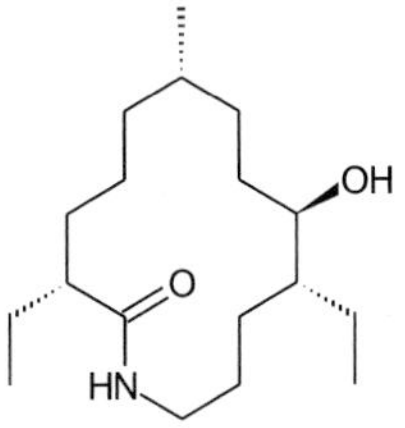

16

Fluvirucin-B_1-Aglycone: Hoveyda (1995)

16.1 Introduction

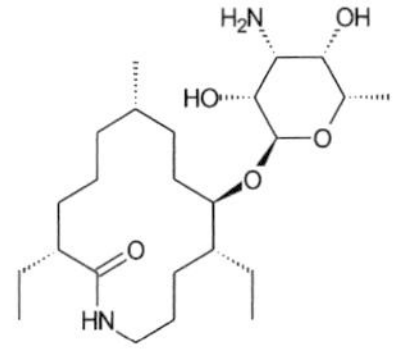

Because of their biological activity, the fluvirucins B_1 (**2**) and B_2 (**3**) represent interesting targets for the synthetic organic chemist. These are macrocyclic lactams that have been used as fungicides and also to combat the influenza-A virus.[1,2] One of the challenges in preparing substances of this type is the need for high selectivity in constructing the stereogenic centers. Moreover, the synthesis should be as flexible as possible in order to facilitate selective preparation of various analogs (including of course ones not found in nature) with the goal of testing for physiological activity.

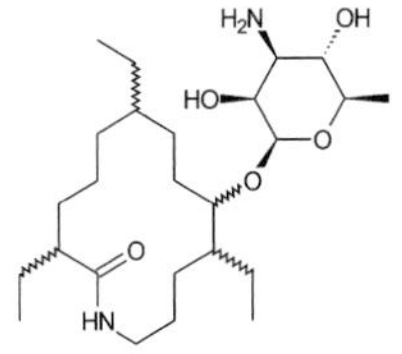

In many cases, transition-metal catalyzed reactions fulfill the criteria cited. For this reason *Hoveda* took considerable advantage of such chemistry in his total synthesis of fluvirucin-B_1-aglycone (**1**).[2]

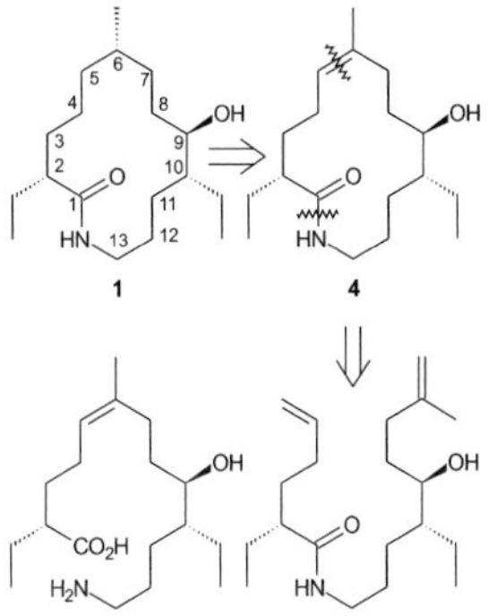

Compound **1** was first analyzed its possible preparation from of alkene **4**. Further retrosynthetic dismembering was directed toward construction of the 14-membered macrocycle. Ring closure via macrolactamization represented one possibility, but that would have required selective construction of the trisubstituted alkene **6**. *Hoveda* regarded this synthesis as sufficiently difficult that he instead utilized an alternative cyclization. This led him to compound **5**.[2] The precise nature of the cyclization and the pathway to the enantiomerically and diastereomerically pure precursor molecule **5** is the subject of this chapter.

16.2 Overview

OH

7

1.

2.

OH

8

OH

9

1.

2. 5 mol% CuBr·SMe_2,
10, THF, **40%**

3. Na, NH_3, -50 °C, **99%**

Ts
N

10

OH

H_2N

11

12

EtMgBr, THF,
(*S*)-[EBTHI]-Zr-binol (**16**)

65%

13

1. ^{n}PrMgBr, THF
3 mol% Cp_2TiCl_2

2.

OH

14

OH

O

15

O
O
Zr

(*S*)-[EBTHI]-Zr-binol

16

15

11

1.

2.

17

18

1.

2.

1
fluvirucin-B 1-aglycone

16.3 Synthesis

Problem

1.
2.
OH
OH
7
8

Tips

- The product is an allylic alcohol. Which of the following methods might be considered for construction of an allylic alcohol: a) addition of a vinyl *Grignard* reagent to an aldehyde; b) addition of an organometallic reagent to an α,β-unsaturated carbonyl compound; or c) reduction of an α,β-unsaturated carbonyl compound?
- An organometallic reagent is in fact added to an α,β-unsaturated aldehyde. What metal should be employed: a) Li; b) Mg; or c) Cu?
- In order to prepare the organolithium compound in the second step, the OH group must be substituted in the first step.

Solution

I
19
O
H
20

The starting alcohol is first transformed into iodide **19** under conditions comparable to those of the *Appel* reaction.[3] Subsequent halogen-metal exchange leads to an organolithium reagent, which subsequently undergoes 1,2-addition with acrolein (**20**). A 1,2-addition reaction with α,β-unsaturated carbonyl compounds is a general property of alkyllithium reagents. If one instead wishes to obtain 1,4-addition it would be necessary instead to utilize a cuprate.

1. Ph_3P, I_2, NEt_3, CH_2Cl_2, 0 °C → 22 °C.
2. ^{t}BuLi, THF, **20**, −78 °C; 51%.

Problem

OH
OH
9
8
9

Tips

- The yield in this step cannot exceed 50%.
- The goal is kinetic resolution of a racemic mixture.
- The method selected involves an asymmetric epoxidation.

Solution

Compound **8** is a secondary allylic alcohol. Systems of this type are well suited to the application of *Sharpless* epoxidation [4] for kinetic racemate resolution,[5] because the chiral titanium reagent bonds directly with the stereogenic center (C-9) in the transition state (**21**). The undesired enantiomer epoxidizes more rapidly than **9**, so **9** can be separated in enantiomerically pure form (see Chapter 6).
$Ti(O^iPr)4$, ^tBuOOH, (+)-diethyl tartrate, MS 4 Å, 33%, >99% *ee*.

9, Ti*, tBu

21

Problem

1.
2. 5 mol% CuBr·SMe_2, **10**, THF, **40%**
3. Na, NH_3, -50 °C, **99%**

OH, Ts, N, H_2N, OH

9 10 11

Tips

- A nitrogen atom is introduced into the molecule in the second step.
- Aziridine **10** is opened like an epoxide using a nucleophile.
- The nitrogen atom of aziridine **10** bears a tosyl group, whereas in **11** it is present as a free amine function.
- The last step is simply cleavage of the tosyl group.
- Attack of a carbon nucleophile on aziridine **10** in the second step completes the carbon skeleton of **11**.
- Addition to the double bond in the first step generates the nucleophile.
- This step also accomplishes introduction of an ethyl group.
- A *Grignard* reagent is added to the double bond.
- Reactions in which an organometallic agent adds to a double bond (e.g., $R'–MgX + R_2C{=}CR_2 \rightarrow R'R_2C\text{-}CR_2MgX$) are called carbometallations. These are catalyzed by transition-metal species.

Solution

A carbometallation is first carried out with ethylmagnesium bromide under zirconocene dichloride catalysis.[6] After transmetallation, the resulting cuprate **23** acts as a nucleophile to open aziridine **10**, leading to compound **24**, which is reduced to **11** with sodium in liquid ammonia.

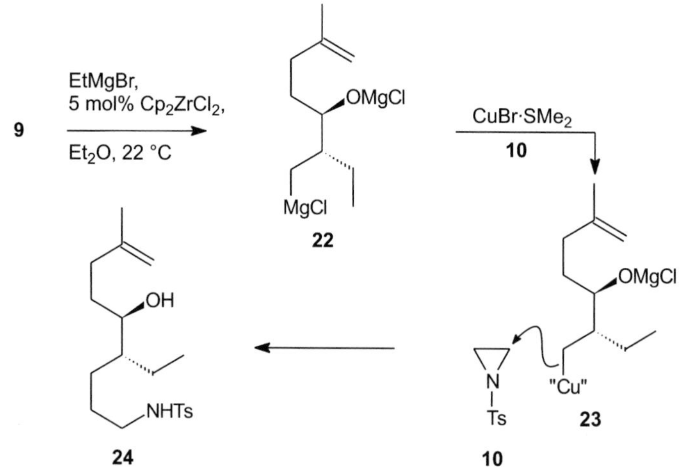

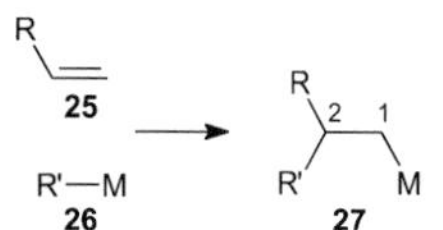

Zirconocene dichloride and related complexes catalyze the addition of trialkylaluminum compounds to terminal acetylenes[7] and olefins[8] as well as the addition of ethyl, propyl, and butyl *Grignard* reagents to alkenes.[9–12] These transformations occur with a high degree of regioselectivity (cf. hydroboration), such that the metal atom always adds to the end position of a terminal multiple bond like that in **25**. If addition is to a triple bond, M and R′ in the product are *cis* to each other.

The mechanism of addition of alkylmagnesium halides to double bonds has been the subject of intensive investigation.[6] According to the current interpretation, these reactions differ significantly in mechanism from reactions with organoaluminum species under similar reaction conditions.[7] For this reason no general mechanism for can be formulated carbometallation reactions.

In the case of the *Grignard* reagent, reaction first occurs between **28** and zirconocene dichloride, leading to zirconium-alkene complex **29** or the zirconacyclopropane **30**. To this is then coordinated alkene **25**, and a further reaction occurs to produce zirconacyclopentane **32**. Here the Zr-C bond forms at the least sterically demanding carbon atom of the alkene.[6] Ethylmagnesium chloride then reacts with **32** to give *at* complex **33**. After transmetallation from zirconium to magnesium, **34** decomposes to product **35** and catalyst **29**.[6]

In order to explain why in **33** the bond from zirconium to C-1 is selectively cleaved rather than the zirconium-C-4 bond it is necessary to take note that this reaction involves allylic and homoallylic ethers and alkoxides. With such species the magnesium cation also coordinates with an oxygen atom from the substrate, as in model system **36**. It can further be shown that in the transition from **33** to **34** zirconium generally remains with the sterically least hindered residue even if there is no heteroatom present.[12]

Ethylmagnesium halide addition can be useful, as in the present total synthesis, for establishing a new stereogenic center. However, this will occur selectively only if a stereogenic center is already present in the allylic position, and in products **40** and **42** one finds an *anti* relationship between the substituents. The only exception is allylic alcohols, which lead to a *syn* arrangement in **38**.[6] It should be noted that **22**, **24**, and **11** have not been presented in a zigzag notation, in contrast to **38**, **40**, and **42**.

5 mol% Cp_2ZrCl_2, EtMgBr, Et_2O, 22 °C, 95:5 *dr.*

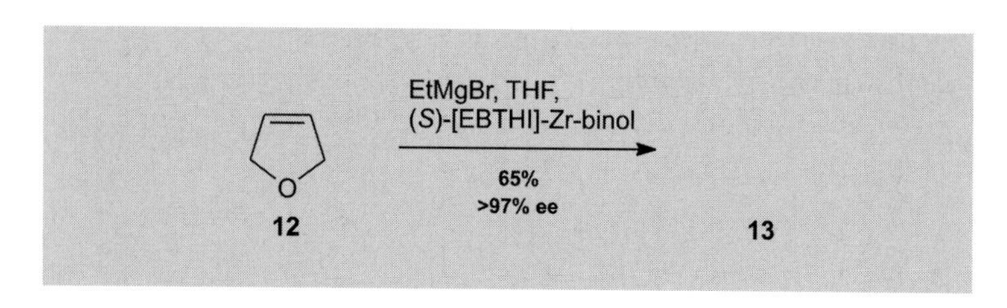

Problem

(*S*)-[EBTHI]-Zr-binol
16

EtMgBr, THF, (*S*)-[EBTHI]-Zr-binol
65%
>97% ee
12 → **13**

Tips

- A similar catalyst has already been employed once in this synthesis.
- The reaction is again a carbometallation.
- The product of carbometallation of the double bond is an organometallic species with an oxygen in the α-position.
- A fragmentation occurs after the carbometallation.

Solution

Et MgBr
43
Et OMgBr
44

Carbometallation of the double bond occurs first. Intermediate **43** decomposes spontaneously in a fragmentation reaction to give alkoxide **44**,[10,11] which leads after protolysis to **13** with >97% *ee*. Concerning the catalytic cycle, catalyst **16** reacts like zirconocene dichloride; i.e., in the first step the binaphthol ligand undergoes substitution by ethylmagnesium bromide. The EBTHI ligand [EBTHI ≡ ethylenebis(tetrahydroindenyl)] is responsible here for chiral induction. A prerequisite for success with this synthetic method is that the terminal double bond in **13** or **44** be carbometallated more slowly than that of starting material **12**.[10]

Problem

13 OH
1. ^{n}PrMgBr, THF
3 mol% Cp_2TiCl_2,
2.
14 OH

Tips

- The overall reaction is a hydrovinylation.
- The process can be viewed as involving two steps, wherein the second step is a cross-coupling reaction with vinyl bromide.
- A transmetallation from magnesium to titanium occurs first.
- The organomagnesium compound contains β-hydrogen atoms.
- Propene is next eliminated, leaving behind a titanium hydride species that adds to the double bond.

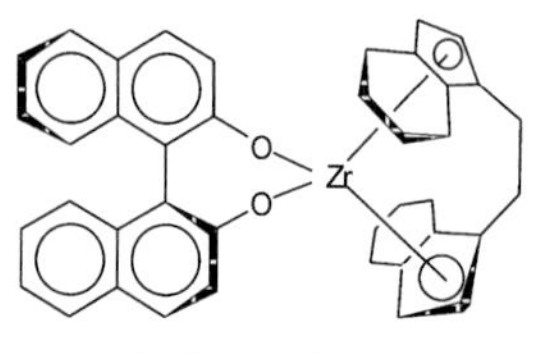

- The reaction is catalytic with respect to titanium.
- What transition metal catalyzes coupling of the *Grignard* reagent with vinyl bromide in the second step: a) Pd; b) Hg; or c) Ni?

Solution

In a formal sense the first step is addition of H-MgX to a double bond. What results is intermediate **45**.[13] Thereafter a *Grignard* cross-coupling reaction is carried out with Ni(0) catalysis.[14]

Addition of H-MgX to the double bond takes place according to the catalytic cycle shown below. A chloride ion from the titanocene dichloride first undergoes substitution by the *Grignard* reagent. Since the alkyl residue contains β-hydrogen atoms, propene (**50**) is then eliminated. Alkene **52** coordinates with **51**, whereupon insertion into the Ti-H bond occurs. A subsequent expulsion regenerates catalyst **49** and releases product **45**.[13]

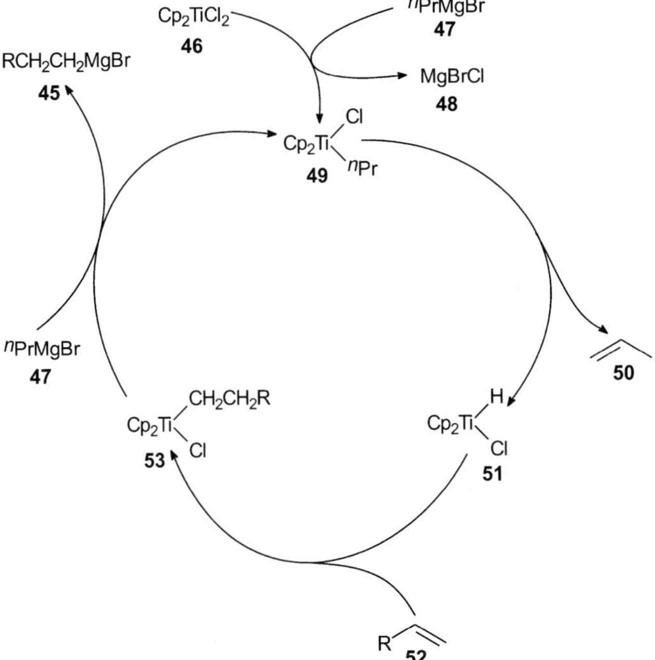

Cross-coupling reactions of organometallic species with vinylic or aryl halides[14] have already been described elsewhere in this book (Chapter 12).

Vinyl bromide, 3 mol% $(Ph_3P)_2NiCl_2$, THF, 72%.

Problem

Tips

- The product is an α-chiral carboxylic acid.
- For this reason the reaction conditions should be kept as pH-neutral as possible.
- The oxidation procedure used here is also applicable to the preparation of aldehydes from primary alcohols.
- The oxidizing agent contains ruthenium.

Solution

In this case TPAP is used for oxidation of the alcohol to a carboxylic acid. Use of TPAP is more common for oxidation to the aldehyde stage (see Chapter 5), but with a longer reaction time it is also possible to cause more complete oxidation to an acid. Since this oxidation takes place under almost neutral conditions, the risk of racemization is minimal relative to that with the alternative chromium-containing oxidizing agents.[15]
5 mol% $(^nPr)_4NRuO_4$ (TPAP), CH_3CN, NMO, 65%.

Problem

Tips

- Two transformations are accomplished here. In what order do they take place?
- An α-chiral carboxylic acid derivative is again involved. An additive is introduced in the first step to ensure that there will be no racemization during formation of the amide.
- The most reactive of the silylating reagents is used for TBS-protection of the alcohol.

Solution

In the first step amine **11** is coupled with carboxylic acid **15** to form an amide. The method employed here for coupling an α-chiral carboxylic acid with an amine was developed in the context of peptide synthesis. Its success is based on DCC-mediated formation (see Chapter 5) of the reactive 1-hydroxybenzotriazole ester **55**, which reacts with an amine to give the corresponding amide. In most cases reaction takes place without racemization, and often in the absence of side reactions that cause other procedures to fail.[16] The alcohol is converted into a silyl ether in the second step.

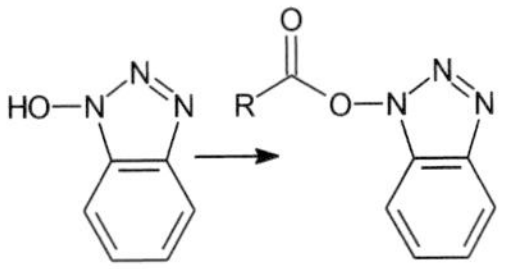

1. DCC, NMM, HOBT (**54**), 22 °C, 60%.
2. TBSOTf, 2,6-lutidine, CH_2Cl_2, 0 °C, 90%.

HO–N N N R O O–N N N

54 **55**

Problem

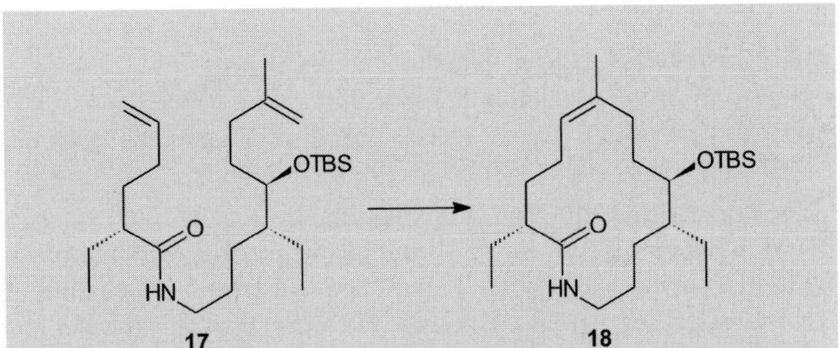

OTBS O HN **17** → OTBS O HN **18**

Tips

- Ethylene is eliminated from the molecule.
- The starting materials for this transformation are two olefins. Two olefins also constitute the products, which are assembled from fragments of the starting materials.
- A specific transition-metal species is capable of catalyzing such transformations.

Solution

The reaction here is an olefin metathesis[17] (for the mechanism see Chapter 5). It is worth noting that the authors attempted to use the most frequently employed catalyst in their synthesis (**56**[17]), but achieved with it less than a 2% yield of cyclic product. Only with introduction of the more reactive molybdenum catalyst **57** did the macrocyclization to **18** take place to give >98% of the product with a double bond in the *Z* configuration.[2]
25 mol% **57**, benzene, 50 °C, 60%.

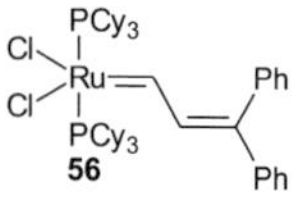

56

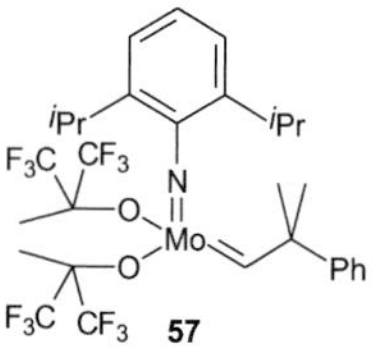

57

Problem

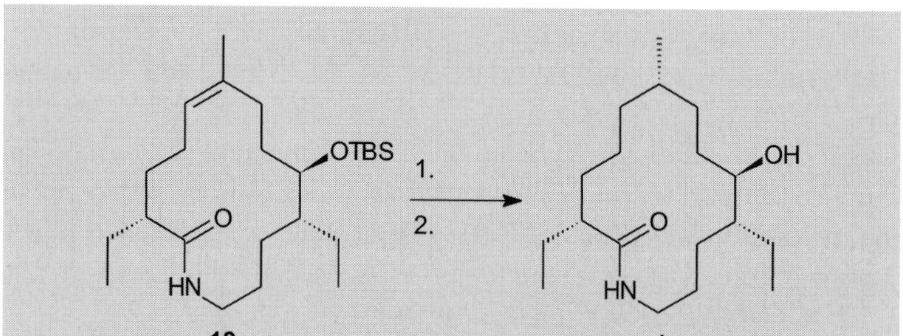

18

1
fluvirucin-B$_1$-aglycone

Tips

- The first step is hydrogenation of the double bond.
- Removal of the protecting group occurs in the last step.
- For this purpose a reagent is used that also decomposes glass.

Solution

Catalytic hydrogenation of the double bond is carried out first, followed by cleavage of the TBS protecting group with hydrogen fluoride in acetonitrile. In an attempt to predict from which side of the molecule H_2 would be transferred during hydrogenation the authors carried out in advance a conformational analysis of **18** with the aid of force-field calculations.[2] Experiment verified the predictions. The selectivity achieved in constructing the last stereogenic center proved to be >95:5 *dr*.

1. H_2, 10% Pd/C, 75%, >95:5 *dr*.
2. HF, CH_3CN, 80%.

16.4 Summary

Hoveyda's synthesis of fluvirucin-B1-aglycone (**1**) presented here is short and convergent. It takes extensive advantage of modern synthetic methods, with transition-metal catalysts or mediators utilized in nine of the fifteen steps. Important key steps worth noting include a zirconocene-catalyzed carbometallation and the olefin metathesis step.

16.5 References

1 Z. Xu, C.W. Johannes, S.S. Salman, A.H. Hoveyda, *J. Am. Chem. Soc.* **1996**, *118*, 10926.

2 A.F. Houri, Z. Xu, D.A. Cogan, A.H. Hoveyda, *J. Am. Chem. Soc.* **1995**, *117*, 2943; H.-G. Schmalz, *Angew. Chem. Int. Ed. Engl.* **1995**, *34,* 1833; *Angew. Chem. Int. Ed. Engl.* **1995**, *34*, 1833; B.M. Trost, M.A. Ceschi, B. König, *Angew. Chem. Int. Ed. Engl.* **1997**, *36,* 1486; *Angew. Chem. Int. Ed. Engl.* **1997**, *36*, 1486; Z. Xu, C.W. Johannes, A.F. Houri, D.S. La, D.A. Cogan, G.E. Hofilena, A.H. Hoveyda, *J. Am. Chem. Soc.* **1997**, *119*, 10302.

3 J.D. Slagle, T.T.-S. Huang, B. Franzus, *J. Org. Chem.* **1981**, *46*, 3526; R. Appel, *Angew. Chem. Int. Ed. Engl.* **1975**, *14,* 801; *Angew. Chem. Int. Ed. Engl.* **1975**, *14*, 801.

4 see also Chapter 6.

5 F.A. Carey, R.J. Sundberg, Advanced *Organic Chemistry*, VCH, Weinheim 1995; R. Brückner, *Reaktionsmechanismen*, Spektrum Akademischer Verlag, Heidelberg 1996, p. 103.

6 A.F. Houri, M.T. Didiuk, Z. Xu, N.R. Horan, A.H. Hoveyda, *J. Am. Chem. Soc.* **1993**, *115*, 6614.

7 E.-I. Negishi, D.Y. Kondakov, D. Choueiry, K. Kasai, T. Takahashi, *J. Am. Chem. Soc.* **1996**, *118*, 9577.

8 D.Y. Kondakov, E.-I. Negishi, *J. Am. Chem. Soc.* **1996**, *118*, 1577; D.Y. Kondakov, E.-I. Negishi, *J. Am. Chem. Soc.* **1995**, *117*, 10771.

9 A.H. Hoveyda, J.P. Morken, *J. Org. Chem.* **1993**, *58*, 4237.

10 J.P. Morken, M.T. Didiuk, A.H. Hoveyda, *J. Am. Chem. Soc.* **1993**, *115*, 6997; M.T. Didiuk, C.W. Johannes, J.P. Morken, A.H. Hoveyda, *J. Am. Chem. Soc.* **1995**, *117*, 7097; A.H. Hoveyda, J.P. Morken, *Angew. Chem. Int. Ed. Engl.* **1996**, *35,* 1262; *Angew. Chem. Int. Ed. Engl.* **1996**, *35*, 1262; M.S. Visser, N.M. Heron, M.T. Didiuk, J.F. Sagal, A.H. Hoveyda, *J. Am. Chem. Soc.* **1996**, *118*, 4291.

11 N. Suzuki, D.Y. Kondakov, T. Takahashi, *J. Am. Chem. Soc.* **1993**, *115*, 8485.

12 T. Takahashi, T. Seki, Y. Nitto, M. Saburi, C.J. Rousset, E.-I. Negislu, *J. Am. Chem. Soc.* **1991**, *113*, 6266.

13 F. Sato, *J. Organomet. Chem.* **1985**, *285*, 53; N. Krause, *Metallorganische Chemie*, Spektrum Akademischer Verlag, Heidelberg 1996, p. 63.

14 L. S. Hegedus, *Transition metals in the synthesis of complex organic molecules*, Wiley-VCH, Weinheim 1995; K. Tamao et al., *Bull. Chem. Soc. Jpn.* **1976**, *49*, 1958; T. Hayashi et al., *J. Org. Chem.* **1986**, *51*, 3772.

15 S.V. Ley, J. Norman, W.P. Griffith, S.P. Marsden, *Synthesis* **1994**, 639.

16 W. König, R. Geiger, *Chem. Ber.* **1970**, *103*, 788.

17 E.L. Dias, S.T. Nguyen, R.H. Grubbs, *J. Am. Chem. Soc.* **1997**, *119*, 3887; M. Schuster, S. Blechert, *Angew. Chem. Int. Ed. Engl.* **1997**, *36,* 2036; *Angew. Chem. Int. Ed. Engl.* **1997**, *36*, 2036.

Abbreviations

Ac	acetyl
AcCl	acetyl chloride
AD	asymmetric dihydroxylation
AIBN	2,2′-azobisisobutyronitrile
tert-AmOH	*tert*-amyl alcohol = 2-methyl-2-butanol
ATP	adenosinetriphosphate
Ar	aryl
BBEDA	(*N,N*′)-bis(benzylidene)ethylene diamine
BINAP	2,2′-bis(diphenylphosphino)-1,1′-binaphthyl
(R)-, (S)-Binol	1,1′-bis-2-naphthol
Bn	benzyl
Boc	*tert*-butoxycarbonyl
BOP	benzotriazol-1-yl-oxytris(dimethylamino)phosphonium hexafluorophosphate
*n*Bu	*n*-butyl
*t*Bu	*tert*-butyl
Bz	benzoyl
CAN	ceric(IV)-ammonium nitrate
cat	catalytic
Cp	cyclopentadienyl
CSA	(+)-camphor-10-sulfonic acid
Cy	cyclohexane
dba	dibenzylideneacetone
DBU	1,8-diazabicyclo[5.4.0]undec-7-ene

DCC	1,3-dicyclohexylcarbodiimide
DDQ	2,3-dichloro-5,6-dicyano-1,4-benzoquinone
DEAD	diethylazodicarboxylate or azodicarboxylic acid diethylester
DET	diethyl tartrate
DHP	2,3-dihydropyran
DHQ	dihydroquinine
DHQD	dihydroquinidine
DIBAH	diisobutylaluminium hydride
DMAP	4-dimethylaminopyridine
DME	1,2-dimethoxyethane
DMF	*N*,*N*-dimethylformamide
DMPU	1,3-dimethyl-3,4,5,6-tetrahydro-2(1H)-pyrimidinone
DMSO	dimethylsulfoxide
DMT	4,4′-dimethoxytriphenylmethyl chloride
dppp	1,3-bis-(diphenylphosphino)-propane
dr	diastereomeric ratio
EBTHI	ethylenebis(tetrahydroindenyl)
ee	enantiomeric excess
Et	ethyl
Et_2O	diethyl ether
HMPA	hexamethylphosphorsäuretriamide
HOAc	acetic acid
HOBT	1-hydroxy-benzotriazole
HOMO	highest occupied molecular orbital
IBD	(diacetoxyiodo)-benzene
ImH	imidazole
(+)-, (–)-Ipc	isopinocamphorylborane
IPOTMS	isopropenyloxytrimethylsilane
KHMDS	potassium hexamethyldisilazane
LAH	lithium aluminium hydride

LDA	lithium diisopropylamide
lk	like
LUMO	lowest occupied molecular orbital
M	metal
MCPBA	*m*-chloroperbenzoic acid
Me	methyl
MOM	methoxymethyl
MOPH	MoO_5*pyridine*HMPA
Ms	methanesulfonyl
MsCl	methanesulfonyl chloride
MS	molecular sieve
NADH	nicotinamide adenine dinucleotide (reduced)
NADPH	nicotinamide adenine dinucleotide phosphate (reduced)
NaHMDS	sodium hexamethyldisilazane
NBS	*N*-bromosuccinimide
NCS	*N*-chlorosuccinimide
NMM	4-methyl-morpholine
NMO	*N*-methylmorpholine-*N*-oxide
PCC	pyridinium chlorochromate
Pd/C	palladium on activated carbon
PDC	pyridinium dichromate
Ph	phenyl
PHAL	phthalazine
PLE	pig liver esterase
PMB	*p*-methoxybenzyl
PMPh	*p*-methoxyphenyl
PPA	polyphosphoric acid
PPTS	pyridinium *p*-toluenesulfonate
*i*Pr	isopropyl
*n*Pr	*n*-propyl

PTSA	*p*-toluenesulfonic acid
Py	pyridine
R	residue
Red-Al®	sodium bis(2-methoxyethoxy)aluminium hydride
RT	room temperature
SDMA	Red-Al®
SET	single electron transfer
SMEAH	Red-Al®
SRS	selfregeneration of stereocenters
TBAF	tetrabutylammonium fluoride
TBAI	tetrabutylammonium iodide
TBS	*tert*-butyldimethylsilyl
TBSCl	*tert*-butyldimethylsilyl chloride
Tf	trifluoromethanesulfonyl
TFA	trifluoroacetic acid
TFAA	trifluoroacetic anhydride
THF	tetrahydrofuran
THP	tetrahydropyran
TMS	trimethylsilyl
TMSCN	trimethylsilyl cyanide
TMSOTf	trimethylsilyltrifluoromethanesulfonate
TNT	2,4,6-trinitrotoluene
*p*Tol	*p*-tolyl
TPAP	tetrapropylammonium perruthenate
TPS	*tert*-butyldiphenylsilyl
Ts	*p*-toluenesulfonyl
TsCl	*p*-toluenesulfonyl chloride
p-TsOH (PTS)	*p*-toluenesulfonic acid
X	halogen atom

Index